Anthony Woodward

The bibliography of the Foraminifera recent and fossil,

Including Eozoon and Receptaculites. 1565 - Jan. 1, 1886

Anthony Woodward

The bibliography of the Foraminifera recent and fossil,
Including Eozoon and Receptaculites. 1565 - Jan. 1, 1886

ISBN/EAN: 9783337717421

Printed in Europe, USA, Canada, Australia, Japan

Cover: Foto ©ninafisch / pixelio.de

More available books at **www.hansebooks.com**

GEOLOGICAL AND NATURAL HISTORY SURVEY OF MINNESOTA.

N. H. WINCHELL, STATE GEOLOGIST.

THE BIBLIOGRAPHY

OF THE

FORAMINIFERA

RECENT AND FOSSIL,

INCLUDING EOZOON AND RECEPTACULITES,

1565 — JAN. 1, 1886.

By ANTHONY WOODWARD.

Part VI of the annual report of progress for the year 1885.

ST. PAUL:
J. W. CUNNINGHAM & CO., STATE PRINTERS,
16 West Fourth Street.
1886.

Note.—This paper is introductory to a contemplated work on the foraminifera and other microscopic organisms of the Cretaceous of Minnesota. According to present plans this work will be done by Messrs. Woodward and Thomas, jointly, and it will be published in one of the volumes of the final report of the survey.

N. H. W.

PREFACE.

This bibliography, which is the result of five years of research, is based largely on facilities afforded by the libraries of the American Museum of Natural History and of the New York Academy of Sciences,—facilities which are not enjoyed by many scientific students. At the beginning I had no idea of presenting this work to the scientific world. When I began the study of the foraminifera I had no knowledge whatever of that which had been done in this branch of science. After I commenced looking up the subject the references accumulated so rapidly that I thought it might be well to collect and put them in shape so that they might be useful to others as well as to myself.

After three years' labor I applied to Mr. H. B. Brady, F. R. S., for information pertaining to the subject. He at once informed me that he had in press a bibliography of the same character, and kindly offered to give me any assistance he could.

When the British Association for the Advancement of Science met at Montreal, in 1884, I met in New York Mr. James Thomson, F. G. S., a member from Scotland, to whom I spoke of my work, asking his advice about proceeding with it. He urged me to continue, and to finish it, as it would become accessible to a great number of workers who could not possess the valuable monograph of Mr. H. B. Brady.

I do not presume that this list is complete; since titles are liable to be found in obscure publications that have not fallen under my notice. Some of those that are here listed may at first appear not to pertain to the subject, but many of the discussions, criticisms, notes, etc., to which reference has been made, although some of them are in general works on microscopy, are of much interest and value to the student, and will be found useful to those who have not access to large libraries.

The list is divided under the following heads:

(1)—Eozoon. (2)—North and South America, Bermuda, Leeward and Windward Islands. (3)—England, Ireland, Scotland and Wales. (4)—France and Italy. (5)—Austro-Hungary, Belgium, Denmark, Finland, Germany, Holland, Netherlands, Norway, Sweden, Switzerland. (6)—Russia and Turkey. (7)—Africa and Asia. The authors' names are then arranged alphabetically and their works according to the date of publication.

I must ask those who may notice omissions or detect errors, to kindly inform me of the same so that I may be able to make corrections in a completed supplement.

I am under great obligations, and return my sincere thanks to the following gentlemen who have rendered me invaluable assistance in sending manuscript lists of their papers.

Rev. P. B. Brodie, M. A., F. G. S., R. V., Warwich, England; Dr. R. Haensler, Sussex, England; Prof. W. C. Williamson, Manchester, England; H. J. Carter, F.R. S., Budleigh, Salterton Devon, England; Joseph Wright, F. G. S., Belfast, Ireland; Sir J. W. Dawson, Montreal, Canada; R. J. Lachmere Guppy, F. L. S., F. G. S., Trinidad; M. O. Terquem, Paris, France; Dr. A. Schneider, Breslau; Prof. Dr. H. B. Geinitz, Dresden, Germany; Prof. Dr. Leopold Auerbach, Breslau, Germany; Prof. Hertwig, Bonn, Germany; Prof. Dr. Carl W. Gumbel, Munich, Germany; Prof. Dr. Haeckel, Jena, Germany; Prof. Dr. Valerian Mœller, St. Petersburg, Russia.

It is hoped that this bibliography will be of some service to the student. The writer will then feel that his years of tedious and constant labor have been well repaid.

ANTHONY WOODWARD.

New York, March 1, 1886.

CONTENTS.

I. Eozoon.
II. North and South America, Bermuda, Leeward and Windward Islands.
III. England, Ireland, Scotland and Wales.
IV. France and Italy.
V. Austro-Hungary, Belgium, Denmark, Finland, Germany, Holland, Netherlands, Norway, Sweden, Switzerland.
VI. Russia and Turkey.
VII. Africa and Asia.

PART I.

EOZOON.

EOZOON.

ANON. Anorganischer Ursprung des Eozoon. <*Verhandl. d. geol k. k. Reichsant.*, p. 58. 1872.

ANON. Eozoon canadense. <*Journ. Roy. Micr. Soc. Lond*, vol. ii. pp. 275, 276, 744, 745, 902. 1879.

ANON. Eozoon canadense. <*Journ. Roy. Micr. Soc. Lond.*, vol. iii., pp. 471, 472. 1880.

BARKER, A. E. Latest observations on Eozoon canadense by Prof. Max Schultze. <*Ann., and Mag. Nat. Hist.* ser. 4, vol. xiii, pp 379-380. 1874.
> Publishing a letter from Prof. Max. Schultze, in which he expresses the opinion that "proper wall" of *Eozoon* is of inorganic origin (Nicholson in White and Nicholson's Bib. p. 75.)

BIGSBY, J. J. On the Laurentian Formation: its mineral constitution, its geographical distribution, and its residuary elements of life. <*Geol. Mag.* Dec. 1, vol. i, pp. 154-158, 200-206. 1864.
> Contains remarks on the distribution of phosphate of lime and carbon in the Laurentian Rocks, and on the occurrence of Eozoon. (Nicholson in White and Nicholson's Bib. p. 77.)

BURBANK, L. S. On Eozoon canadense in the crystalline Limestones of Massachusetts. <*Amer. Nat.*, vol. v, pp. 535-539. 1871.

BURBANK, L. S. On Eozoon canadense in the crystalline Limestones of Massachusetts. <*Proc. Am. Assoc. Adv. Sci.*, 1871, vol xx, pp. 262-266. 1872.

BURBANK, L. S. Views on the *Eozoonal* limestones of Eastern Massachusetts. <*Proc. Bost. Soc. Nat. Hist.*, vol. xiv, pp. 194-198. 1872.

CARPENTER, W. B. On the Structure and Affinities of *Eozoon canadense*. *Proc. Roy. Soc.*, vol. xiii, pp. 545-549. 1860.

CARPENTER, W. B. Additional Note on the Structure and Affinities of Eozoon canadense. <*Quart. Journ. Geol. Soc. Lond.*, vol. xxi, pp. 59-66, 2 plates and wood cuts. 1865.

CARPENTER, W. B. On the Structure, Affinities, and Geological Position of Eozoon canadense. <*Intellectual Observer*, No. xl, p. 278, 2 plates. 1865.

CARPENTER, W. B. Eozoon canadense. *Intellectual Observer*, No xl, p. 300. 1865.

CARPENTER, W. B. Supplemental Notes on the Structure and Affinities of Eozoon canadense. *Quart. Journ. Geol. Soc. Lond.*, vol. xxii, pp. 219-228. 1866.

CARPENTER, W. B. Notes on the Structure and Affinities of Eozoon canadense. <*Canad. Nat.*, new ser., vol. ii, pp. 111-119, wood cut. 1865. A reprint from *Quart. Journ. Geol. Soc. Lond.*, 1865. (Nicholson in White and Nicholson's Bib. p. 87.)

CARPENTER, W. B. Further Observations on the Structure and Affinities of Eozoon canadense. In a letter to the president. <Proc. Roy. Soc. Lond., vol. xxv, pp. 503-508. 1867.
> A resume of the state of the Eozoon controversy at the time—1867. (Nicholson in White and Nicholson's Bib. p. 87.

CARPENTER, W. B. On the Eozoon canadense. <Nature, vol. iii, pp. 185, 186, 386. 1871.

CARPENTER, W. B. New Observation on Eozoon canadense. <Ann., and Mag. Nat. Hist., ser. 4, vol. xiii, pp. 456-470, 1 plate. 1874.
> The author treats more especially of the nummulinæ layer and the canal-system of the "intermediate skeleton," and concludes by summarizing the general evidence in favor of the organic origin of Eozoon. (Nicholson in White and Nicholson's Bib. p. 87.)

CARPENTER, W. B. Final Note on Eozoon canadense. <Ann., and Mag. Nat. Hist., ser. 4, vol xiv, pp. 371-372. 1874.

CARPENTER, W. B. Remarks on Mr. H. J. Carter's Letter to Prof. King on the Structure of the so-called Eozoon canadense. <Ann., and Mag. Nat. Hist., ser. 4, vol. xiii, pp. 277-284 with 2 engravings. 1874.
> A recapitulation of the principal facts in support of the belief that Eozoon canadense is a Foraminifer. (White's Bib. p. 87.)

CARPENTER, W. B. Remarks on Eozoon canadense. <Nature, vol. ix, p. 491. 1874. (Abstract.)
> His reply to Mr. Carter's letter to Prof. King on the structure of the so-called Eozoon canadense.

CARPENTER, W. B. Further Researches on Eozoon canadense. <Nature, vol. x, p. 390. 1874.

CARPENTER, W. B. On the Replacement of Organic Matter by Siliceous Deposits in the process of Fossilization. <Nature, vol. x, p. 452. 1874. (Abstract.)

CARPENTER, W. B. Further Researches on Eozoon canadense. <Rep. Brit. Assoc. for 1874, Section, pp. 136, 137. 1875.

CARPENTER, W. B. New Laurentian Fossil. <Nature, vol. xiv, pp. 8, 9. 1876.

CARPENTER, W. B. Supposed New Laurentian Fossil. <Nature, vol. xiv, p. 68. 1876.

CARPENTER, W. B. Note on Otto Hahn's Microgeological Investigation of Eozoon canadense. <Ann., and Mag. Nat. Hist., ser. 4, vol. xvii, pp. 417-422. 1876.

CARPENTER, W. B. The Eozoon canadense. <Nature, vol. xx, pp. 328-330. 1879.

CARPENTER, W. B. Eozoon canadense. <The Microscope and its Revelations, Sixth Edition, pp. 587-592. 1881.

CARPENTER, W. B., and J. W. Dawson. The Eozoon canadense. <Nature, vol. xx, p. 328. 1879.

CARTER, H. J. On the structure called Eozoon canadense, in the Laurentian Rocks of Canada. <Ann., and Mag. Nat. Hist., ser. 4, vol. xiii, pp. 189-193. 1874.
> Gives reasons for believing that Eozoon is not of organic origin. (Nicholson in White and Nicholson's Bib. p. 88.)

CARTER, H. J. On the structure called Eozoon canadense in the Laurentian Limestones of Canada. *Ann., and Mag. Nat. Hist.*, ser. 4, vol. xiii, pp. 376-378, with 1 engraving. 1874.

CARTER, H. J. Relation of the Canal-system to the Tubulation in the Foraminifera, with reference to Dr. Dawson's "Dawn of Life." *Ann., and Mag. Nat. Hist.*, ser. 4, vol. xvi, pp. 420-424. 1874.

Discusses the minute structure of the test of recent *Foraminifera*, as bearing on the nature of *Eozoon canadense*. (Nicholson in White and Nicholson's Bib. p. 88.)

CARTER, H. J. Eozoon canadense not a Foraminifer or calcareous Rhizopod secretion. <*Amer. Journ. Sci.*, vol. vii, 3d ser., pp. 437, 438. 1874.

CREDNER, H. Die Gliederung der eozoischen (vorsilurischen) Formationsgruppe Nord-Amerikas. <*Zeit. Gesam. Naturwissenschaften*, 32, pp 352-405. 1868.

DANA, J. D. On the History of Eozoon canadense. *Am. Journ. Sci.*, vol. xi, 2d ser., pp. 344-362, wood cuts and 1 plate. 1865.

This article appears in the Journal without a name; *i. e.* editorially. This history embraces a full discussion of the subject, and includes a complete description and illustration of the structure of the fossil, and the chemical composition of specimens. (White in White and Nicholson's Bib. p. 22.

DANA, J. D. Manual of Geology. Second edition, pp. 158, 159. 1875.

D'ARCHIAC. Note sur l'existence des restes organiques dans les Roches Laurentiennes du Canada. <*Comptes Rendus*, vol. liii, pp. 192-194. 1865.

A note presented by M. D'Archiac on the part of Dr. W. B. Carpenter as to the discovery of *Eozoon canadense*. (Nicholson in White and Nicholson's Bib. p. 90.)

DAWSON, J. W. On the Structure of certain Organic Remains in the Laurentian Limestones of Canada. <*Quart. Journ. Geol. Soc. Lond.*, vol xxi, pp. 51-59, pls. vi, vii. 1865.

The author gives a detailed description of the structure of the bodies described by Sir. William Logan as being organic and as occurring in the Lower Laurentian Limestones. (Quart Journ. Geol. Soc , vol. xxi, p. 45.) The generic name of Eozoon is proposed for these, and the single form described is discussed under the name of *Eozoon canadense*. The author further concludes that Eozoon is probably to be regarded as an ancient type of the *Foraminifera*. (Nicholson in White and Nicholson's Bib. p 93.)

DAWSON, J. W., and W. B. CARPENTER. Notes on Fossils recently obtained from the Laurentian Rocks of Canada, and on objections to the Organic nature of Eozoon. *Quart. Journ. Geol. Soc. Lond.*, vol. xxiii, pp. 257-265, 2 plates. 1865.

DAWSON, J. W. Notes on fossils recently obtained from the Laurentian Rocks of Canada, and objections to the Organic nature of Eozoon. <*Amer. Journ. Sci.*, vol. xliv, 2d ser., pp. 367-376. 1867.

The article also contains notes by W. B. Carpenter; and "Summary" and "conclusion" of King and Rowney, on the same subject; the latter gentlemen opposing, and the former advocating, the organic origin of Eozoon. (White in White and Nicholson's Bib. p. 22.)

DAWSON, J. W. On certain Organic remains in the Laurentian Limestone of Canada. *Canad. Nat.*, new ser., vol. 11, pp. 99-111, 127, 128. 3 wood cuts. 1865.

> A reprint from the *Quart. Journ. Geol. Soc. Lond.*, 1865, with some additional notes. A short appendix to the paper follows at pp. 127, 128. (Nicholson in White and Nicholson's Bib. p. 93.)

DAWSON, J. W. Notes on fossils recently obtained from the Laurentian Rocks of Canada, and on objections to the Organic nature of Eozoon, with notes by W. B. Carpenter, M. D., F. R. S. <*Quart. Journ. Geol. Soc. Lond*, vol. xxiii, pp. 257-265, pls. xi, xii. 1867.

> In the first part of this memoir, Dr. Dawson gives an account of the general appearance and microscopic structure of a specimen of *Eozoon canadense*, found in the Laurentian rocks at Tudor, in which the chambers of the skeleton are filled with a dark colored coarse limestone. The author next deals with certain specimens from Long Lake and Wentworth, and also from Madoc, and concludes by reviewing the objections brought forward by Professors King and Rowney to the organic nature of Eozoon. Dr. W. B. Carpenter adds a note on the appearances presented by thin slices of specimens of *Eozoon* in which the canal-system has been infiltrated with transparent carbonate of lime. (Nicholson in White and Nicholson's Bib. p. 93.)

DAWSON, S. W., and W. B. CARPENTER. Notes on Fossils recently obtained from the Laurentian Rocks of Canada, and objections to the organic nature of Eozoon. <*Amer. Journ. Sci.*, vol. xliv, 2d ser., pp. 367-376. 1867.

DAWSON, J. W., and W. B. CARPENTER. On new specimens of Eozoon canadense, with a reply to the objections of Professors King and Rowney. <*Amer. Journ. Sci*, vol. xlvi, 2d ser., pp. 245-255, 2 plates. 1868.

DAWSON, J. W. On new specimens of Eozoon canadense, with a reply to Professors King and Rowney; with notes by W. B. Carpenter. <*Amer. Journ. Sci.*, vol. xlvi, 2d ser., pp. 245-257, 2 plates. 1868.

> The authors advocate the organic origin of Eozoon. (Nicholson in White and Nicholson's Bib. p. 22.)

DAWSON, J. W. Remarks on Eozoon canadense. <*Nature*, vol. x, p. 103. 1 wood cut. 1874.

DAWSON, J. W. Notes on the occurrence of Eozoon canadense, at Cote St. Pierre. <*Nature*, vol. xii, p. 79. 1875. (Abstract.)

DAWSON, J. W. On the Eozoon canadense. <*Nature*, vol. iii, p. 287. 1871.

DAWSON, J. W. The Story of the Earth and Man.
> (Remarks on Eozoon chapter ii, iii, pp. 17-38.) pp. 403, 800. London, 1873.

DAWSON, J. W. The Dawn of Life; being the history of the oldest known fossil remains, and their relations to geological time and to the development of the animal kingdom, pp. 239, with 8 plates and 49 wood cuts. London, 1875.

> This work deals principally with the history of the discovery of *Eozoon canadense*, and with all the known facts bearing on its structure and nature. The author first gives a descriptive sketch of the Laurentian formation, accompanied by sections, and a colored map showing the distribution of the Laurentian Limestones in the counties of Ottawa and Argenteuil. Next, a history is given of the various steps which led to the discovery of *Eozoon*, and a record of its interpretation by Carpenter and the author

Thirdly, a chapter is devoted to a consideration of the minute structure exhibited by *Eozoon;* and this is compared with the structure of recent *Foraminifera*. The fifth chapter is concerned with the manner in which *Eozoon* has been preserved, and with a consideration of the processes of fossilization by infiltration in general. In the sixth chapter, the author deals with the successors and contemporaries of *Eozoon*, with special reference to *Archæosphærina, Stromatopora, Caunopora,* and *Receptaculites.* Another chapter is devoted to a consideration of the various objections which have been urged against the organic nature of *Eozoon;* and a final chapter treats of certain speculative considerations which may be drawn from the study of this fossil. (Nicholson in White and Nicholson's Bib. p. 95.)

DAWSON, J. W. On Mr. Carter's objections to Eozoon. <*Ann., and Mag. Nat. Hist*, ser. 4, vol. xvii, pp. 118, 119. 1876.

DAWSON, J. W. Notes on the Phosphates of the Laurentian and Cambrian Rocks of Canada. <*Quart. Journ. Geol. Soc. Lond.*., vol. xxxii, pp. 285–291. 1876.

> Concludes that the phosphatic material found in these rocks in Canada is of organic origin, and has been produced by the agency of marine invertebrates. (Nicholson in White and Nicholson's Bib. p. 95.)

DAWSON, J. W. Notes on the Occurrence of Eozoon canadense at Cote St. Pierre. <*Quart. Journ. Geol. Soc. Lond.*, vol. xxxii, pp. 66–74, plate x, with 4 wood cuts. 1876.

> The author gives an account of the nature and arrangement of the strata at Cote St. Pierre, with special reference to the appearance presented by *Eozoon* as occurring *in situ.* Numerous chrysotile veins pass through the limestone, but the author concludes that they are altogether subsequent to the fossil in origin. The close resemblance of weathered specimens to *Stromatopora* is insisted upon; and two new forms of *Eozoon canadense* are described as var. *minor* and var. *acervulina.* The limestone sometimes contains numerous little globose casts of chamberlets, single or attached in groups. each of which possesses the structure of the "proper wall" of *Eozoon.* For these the author proposes the name of *Archæosphærinæ.* (Nicholson in White and Nicholson's Bib. pp. 95, 96.)

DAWSON, J. W. On some new specimens of Fossil Protozoa from Canada. <*Proc. Am. Assoc. Adv. Sci.*, vol. xxiv, pp. 100–106, wood cuts. 1876.

> The author gives general description and illustration of *Eozoon canadense,* and also *Foraminifera,* from Cretaceous rocks. He advocates the organic origin of *Eozoon.* (White in White and Nicholson's Bib. p. 22.)

DAWSON, J. W. New Facts relating to Eozoon canadense. <*Proc. Am. Assoc. Adv. Sci.*, vol. xxv, pp. 231–234. 1876.

> The fossil nature of *Eozoon canadense* is advocated. (White in White and Nicholson's Bib. p 22.)

DAWSON, J. W. Eozoon canadense according to Hahn. *Ann. Mag. Nat. Hist.*, ser. 4, vol. xviii, pp. 29–38. 1877.

> A critical notice of a memoir by Hahn (see post.) in which the latter endeavors to show that *Eozoon* is a purely mineral structure. (Nicholson in White and Nicholson's Bib. p. 96.)

DAWSON, J. W. New Facts relating to Eozoon canadense. *Canad. Nat.* new ser., vol. viii, pp. 262–265. 1878.

DAWSON, J. W. On the Microscopic Structure of Stromatoporidæ, and on Palæozic Fossils mineralized with Silicates, in illustration of Eozoon. *Quart. Jour. Geol. Soc Lond.* vol. xxxv, pp. 48–66. 3 plates. 1879.

DAWSON, J. W. Note on recent Controversies respecting Eozoon canadense. *Can. Nat.*, vol. ix; p. 228. 1879.

DAWSON, J. W. Mœbius on Eozoon canadense. <*Amer. Jour. Sci.*, vol. xvii, p. 196, wood cuts. 1879.

DAWSON, J. W. Notes on Eozoon canadense. <*The Can. Rec. of Sci.*, vol. i, pp 58, 59. 1884.
: Abstract of a paper read before the British Association at Southport, 1883.

DAWSON, J. W. On the Geological Relations and Mode of Preservation of Eozoon canadense. <*Report Brit. Assoc.* (Southport, 1883), p. 494. 1884.

DAWSON, J. W. Canadian and Scottish Geology. An Address delivered before the Edinburgh Geological Society at the close of the Session, 1884. *Trans. Edin. Geol. Soc.*, vol. v, pp. 113, 114. 1885.
: Remarks on Eozoon canadense.

EDWARDS, A. M. Microscopical Examination of Two Minerals.* <*Proc. Lyceum Nat. Hist.*, pp. 96-98. 1870.
: [*Supposed to be Eozoon.]

FRITSCH, A. Ueber das Vorkommen des Eozoon im nordlichen Bohmen. *Neues Jahrb. fur Min.*, etc., pp. 352-354. 1866.

FRITSCH, A. Ueber Eozoon bohemicum aus dem Kornigen Kalke von Raspenau in Bohmen. <*Landesdurchforschung von Bohmen,*—Geol. Sect., pp. 245-251, 1 wood cut and 2 plates. 1869.
: Not seen.

FRIC, ANTON (Dr.) Ueber Eozoon bohemicum, Fr., aus den Kornigen Kalkstein von Raspenau bei Friedland in Bohmen. <*Geologie von Bohmen* vol. i, pp. 245-256, 1 wood cut, 2 plates. 1869.

GUMBEL, C. W. Ueber das Vorkommen von Eozoon in dem ostbayerischen Urgebirge. <*Sitzungsber d. k. b. Akad. Wiss. Munch.* 1866, Bd. i, pp. 25-144, 3 plates.

GUMBEL, C. W. Eozoon im ostbayer. Urgebirge. <*N. Jahrb. fur Min.* etc., 1866. I. S. 1 und *N. Jahrb. fur Min.*, etc. 1866. S. 210.

GUMBEL, C. W. Eozoon im Urkalke von Sachsen. <*N. Jahrb. fur Min.*, etc. 1866. S. 579.

GUMBEL. On the Occurrence of Eozoon in the Primary Rocks of Eastern Bavaria. <*Quart. Journ. Geol. Soc. Lond.*, vol. xxii, pp. 23, 24. 1866.
: A review by H. M. J.

GUMBEL, C. W. On the Laurentian Rocks of Bavaria. <*Cand. Nat.* new series, vol. iii, pp. 81-101, 1 plate. 1868 Translated from the proceedings of the Royal Bavarian Academy for 1886, by Prof. Markgraf.
: [*EDITOR'S NOTE.—In revising and preparing this for the press, the original paper has been considerably abridged by the omission of portions, whose place is indicated in the text. Some explanatory notes have also been added.—T. S. H.]

GUMBEL, C. W. Eozoon im Kornigen Kalke Schwedens. <*Leonhard und Geinitz neues Jahrbuch.* 1869. pp. 551-559. 1869.

GUMBEL, C. W. Ueber die Natur von Eozoon. 8 p. Ratisbonne. 1876.
: Not seen; title taken from a catalogue.

HAHN, DR O. Giebt es ein Eozoon canadense? Eine mikrogeologische Untersuchung. ∕ *Wurttembergische naturwiss.* Jahreshefte, 32 Jahr, pp. 132-155. (Translated by W. S. Dallas, Ann. and Mag. Nat Hist, ser. 4, vol. xvii, pp. 265, 282) 1876.

    After an examination of serpentinous limestones from Canada and Europe, the author concludes that *Eozoon canadense* is of organic origin. (Nicholson in White and Nicholson's Bib. p. 103.)

HAHN, DR. O. Giebt es ein Eozoon canadense? Eine mikrogeologische Untersuchung. ∕ *Wurtt. naturwiss.* Jahresheften, 1876, 24 pp. Stuttgart, 1876.

    Printed as a separate pamphlet.

HAHN, (DR.) O. Giebt es ein Eozoon canadense? Erwiderungauf Dr. C. W. Gumbel's und Dr. Carpenter's Entgegnung. ∕ *Wurtt. naturwiss* Jahresheften. Jahrgang. 1878. 21 pp. 1 plate. Stuttgart. 1878.

HAHN, O., in Reutlingen sprach uber das Eophyllum canadense aus dem Serpentinkalk des Laurentiangneisses von Canadas. ∕ *Wurttemb. naturwiss.* Jahresheften. Jahrgang. 1880. pp. 71-74.

HAHN, O. Die Meteorite (chondrite) und ihre Organismen, 56 pp. 32 plates. 1880.

    Plate xxx., fig. 5, Eozoon canadense, reputed canal system of Eozoon; fig. 6 the same. Both stones from which the slides were taken were collected by me in Little Nation. Let one compare the canal system of the nummulite fig. 3 with this reputed canal system! Figs. 3 and 5 should be the same thing.

HALL, J. Note upon the Geological position of the Serpentine Limestone of Northern New York, and an inquiry regarding the relations of this Limestone to Eozoon Limestones of Canada. ∕ *Amer. Journ. Sci.*, vol xii, 3d ser., pp. 298-300, 1876.

    Abstract of the paper read before the Amer. Association at Buffalo.

HAUER, M. Das Eozoon canadense. Eine micro-geologische Studis, mit 18 photographic plates. Leipzig. 1885.

HITCHCOCK, C. H. The Earlier Forms of Life (Eozoon). 16 pp. 10 figs. N. P. N. D.

HOCHSTETTER, —. Eozoon in Austria. *Quart. Journ. Geol. Soc. Lond.*, vol. xxii, p. 16. 1866.

HOCHSTETTER, R. F. Ueber das Vorkommen von Eozoon in Krystallinischen Kalke von Krumman in sudlichen Bohmen. ∕ *Sitz. d. k. Akad. d. Wiss.*, vol liii, pp. 14-25. 1866.

HOCHSTETTER, (Prof V.) DR. W. B, CARPENTER IN LONDON. *Neuer Fund von Eozoon canadense.* K. K. geol. Reich. Ver. 1868, pp. 69, 70. 1868.

HOFFMANN, R. On the Mineralogy of Eozoon canadense. <*Amer. Journ. Sci*, vol. i, 3d ser., pp. 378, 379. 1871.

HUNT, T. S. Laurentian Rhizopods of Canada. (Extract of a letter from T. Sterry Hunt, F. R. S., to J. D. Dana, April 2, 1864.) *Amer. Journ. Sci.*, vol. xxxvii, 2d ser., p. 431. 1864.

HUNT, T. S. On the Mineralogy of Eozoon canadense. <*Canad. Nat.*, n. s., vol. ii, pp. 120-127, 1 plate. 1865.

HUNT, T. S. On the Mineralogy of certain Organic Remains from the Laurentian Rocks of Canada. <*Quart. Journ. Geol. Soc. Lond.*, vol. xxi, pp. 67-71. 1865.

> Gives a detailed account accompanied with analysis, of the mineral nature and structure of *Eozoon Canadense*. (Nicholson in White and Nicholson's Bib. p. 107.)

HUNT, T. S. Geology and mineralogy of the Laurentian Limestones. <*Geological Survey of Canada*. Report of progress from 1863 to 1866, pp. 181-233. Ottawa, 1866.

> Though essentially mineralogical, this report contains many interesting observations bearing on the nature and mode of preservation of *Eozoon canadense*. (Nicholson in White and Nicholson's Bib. p. 107.)

HUNT, T. S. The Geological History of Serpentines, including Notes on pre-Cambrian Rocks. <*Trans. Roy. Soc.* Canada, vol. i, pp. 165-215. 1883.

JONES, T. R. Eozoon canadense in this country. <*Nat. Hist. Rev. Lond.*, vol. v., pp. 297, 298. 1865.

> In this communication to the editor he states that Eozoon canadense is abundant in the British Isles. Mr. W. A. Sanford has hunted it up in the Green Connemara marble, and he also finds it there in masses indicated by him. The best way of getting a sight of the structure due to the presence of Foraminifera is to dissolve small flakes of the "Irish Green" in very weak dilute acid, and then the shelly part being removed, the green silicates remain representing the sarcode that filled the chambers, pseudopodian tubules and stolon passages.

JONES, T. R. On the Oldest Known Fossil, Eozoon canadense. <*Popular Sci. Rev.*, vol. iv., pp. 343, pl. xv. 1865.

> Discusses the geological and zoological relations of *Eozoon*.

JONES, T. R. On the Oldest Known Fossil, Eozoon canadense of the Laurentian Rocks of Canada; its place, structure and significance. <*Popular Sci. Review*, 1867, pp. 343-352, plates xv and 2 wood cuts.

> A semi-popular account of *Eozoon canadense*. (Nicholson in White and Nicholson's Bib. p. 109.)

JULIEN, A. A. A study of "Eozoon Canadense." Field observations. <*Proc. Amer. Asso. Adv. Sci.*, vol. xxxiii, 1884, pp. 415, 416. (Abstract.) 1885.

KING, W. Note on Eozoon canadense. <*Nature*, vol. iv, p. 85. 1871.

KING, W., and T. H. ROWNEY. On the so called "Eozoonal Rock." <*Quart. Journ. Geol. Soc. Lond.*, vol. xxii, pp. 185-218, 2 plates. 1866.

> The authors describe in this memoir the results of a careful chemical and microscopical examination of the Grenville "*Eozoonal*" Ophite, from which they arrive at the conclusion that Eozoon canadense is of truly inorganic origin. (Nicholson in White and Nicholson's Bib. p. 110.

KING W., and T. H. ROWNEY. On the so-called "Eozoonal" Rock. <*Quart. Journ. Geol. Soc. Lond.*, vol. xxv, pp. 116, 117. (Abstract.) 1869.

> The authors adduce further evidence that their views as to the mineral nature of Eozoon are correct. (Nicholson in White and Nicholson's Bib. p. 110.)

KING, W., and T. H. ROWNEY. On Eozoon canadense. <*Proc. Roy. Irish Acad.*, vol. x, p. 506, 2 plates xlii—xliv. 1870.

KING, W., and T. H ROWNEY. On the Mineral Origin of the so-called "Eozoon canadense." <*Proc. Roy. Irish Acad.*, ser. 2, vol. 1, pp. 140-153. 1871.

    A reply to papers by Drs. J. W. Dawson and T. Sterry Hunt on the zoological and chemical aspects of the question respectively. The paper concludes with a recapitulation of the various points detailed in the formerly published papers of the authors. (Nicholson in White and Nicholson's Bib. p. 110.)

KING, W., and T. H. ROWNEY. Eozoon, examined principally from a Foraminiferal standpoint. <*Ann., and Mag. Nat. Hist.*, ser. 4, vol. xiv, pp. 274-289, plate xix. 1874.

    A controversial paper, in which evidence is brought forward to show that *Eozoon canadense* is inorganic in its nature. (Nicholson in White and Nicholson's Bib. p.111.)

KING, W., and T. H. ROWNEY. Remarks on the subject of Eozoon. <*Ann., and Mag. Nat. Hist.*, ser. 4, vol. xiii, pp. 390-396. 1874.

    A summary of the chief points in favor of the mineral nature of *Eozoon canadense*. (Nicholson in White and Nicholson's Bib. p. 111.)

KING, W., and T. H. ROWNEY. Remarks on the "Dawn of Life" by Dr. Dawson; to which is added a supplementary note. <*Ann., and Mag. Nat. Hist.*, ser. 4, vol. xvii, pp. 360-377. 1876.

    A critical memoir, stating the objections held by the authors as to the supposed organic origin of Eozoon. (Nicholson in White and Nicholson's Bib. p. 111.)

LAUBE, G. Notizer von einer Reise in Scandinavien. <*Lotos*, xxiv, Jahrg. (Eozoon p. 21.) Prague. 1874.

LEA, I. Contributions to Geology. Philadelphia, 1833.

LEIDY, J. Remarks on Eozoon. <*Proc. Acad. Nat. Sci.*, 1877, p. 20. 1877.

LOGAN, W. E. Supposed Fossils in the Laurentian Limestone. <*Geology of Canada* pp. 48, 49, 2 wood cuts. 1863.

LOGAN, W. E. On Organic Remains in the Laurentian Rocks of Canada; (from a letter to the editors of this Journal from Sir W. E. Logan, F. R. S, dated Montreal, Feb. 17th, 1864.) <*Amer. Journ. Sci.*, vol. xxxvii, 2d ser., pp. 272, 273. 1864.

LOGAN, W. E. On the Occurrence of Organic Remains in the Laurentian Rocks of Canada. <*Quart. Journ. Geol. Soc. Lond.*, vol. xxi, pp. 45-50. 1865.

    This memoir is a geological one, occupied with a general description of the Laurentian Rocks of Canada, illustrated by sections. The author, however, gives an account of the discovery of *Eozoon* in the Lower Laurentian Limestone, and describes the general mode of occurrence of, and the appearance presented by, the specimens. (Nicholson in White and Nicholson's Bib. p. 112.)

LOGAN, W. E. On the Occurrence of Organic Remains in the Laurentian Rocks of Canada. *Canad. Nat.*, new ser, vol. ii, pp. 92-99. 1865.

    A reprint from the Quart. Journ. Geol. Soc. Lond., 1865, with some additional notes. (Nicholson in White and Nicholson's Bib. p 112.)

LOGAN, W. E. On New specimens of Eozoon. *Quart. Journ. Geol. Soc. Lond.*, vol. xxiii, pp. 253-257. 1867.

    This is a geological memoir, it is of interest to the palæontologist as giving a detailed account of the precise geological position of the bed from which was obtained the least altered example of *Eozoon canadense* (the "Tudor specimen") as yet known to science. (Nicholson in White and Nicholson's Bib. p. 113.)

LOGAN, W. E., J. W. Dawson, and T. S. Hunt. On the Occurrence of Organic Remains in the Laurentian Rocks of Canada. <*Report Brit. Assoc.* (Bath Meeting), Trans. Sections, p. 225. 1864.

MOEBIUS, K. Der Bau des Eozoon canadense nach eigenen Untersuchungen verglichen mit dem Bau der Foraminiferen. <*Palaeontographica*, vol. xxv, pp. 175-192, plates 23-40. 1878.

MOEBIUS, DR. K. Ist das Eozoon ein versteinerter Wurzelfussler oder ein Mineralgemenge? <*Die. Natur.* Jahrg. 1879, Nos. 7, 8, 10-21, wood cuts. 1879.

MOEBIUS, DR. K. Principal, J. W. DAWSON's Criticism of my Memoir "On the Structure of Eozoon canadense compared with that of Foraminifera." <*Amer. Journ. Sci.*, vol. xviii, 3d ser., p. 177. 1879.

NICHOLSON, H. A., and DR. J. MURIE. On the Minute Structure of *Stromatopora* and its Allies. <*Journ. Linn. Soc.*, vol. xiv, pp. 187-246, 5 wood cuts and 5 plates. 1878.

PARKER, JONES, and BRADY. On the Priority in the Discovery of the Canal System in Foraminifera. <*Ann., and Mag. Nat. Hist.*, ser. 4, vol. xiv, p. 64, 305. 1874.

PERRY, J. B. Eozoon Limestone of Eastern Massachusetts. <*Amer. Nat.*, vov. v, pp. 539, 541. 1871.

PERRY, J. B. Notes on Eozoon canadense. <*Nature*, vol. iv, p. 28. 1871.

PERRY, J. B. On "the Eozoon" Limestone of Eastern Massachusetts. <*Proc. Am. Assoc. Adv. Sci.*, vol. xx, 270-276. 1872.

> Mr. Perry corroborates the statement of Mr. Burbank as to the existence of *Eozoon* in the chrystalline limestones of Eastern Massachusetts. (Nicholson in White and Nicholson's Bib. p. 57.)

PERRY, J. B. Few remarks on the "*Eozoon*" Limestone of Eastern Massachusetts. <*Proc. Bost. Soc. Nat. Hist.*, vol. xiv, pp. 199-204. 1872.

PUSYREWSKI, (PROF.) P. Eozoon canadense im Kalkstein von Hopinwara in Finnland. <*Bull. d. l'Acad, Imp, d. Sci. a. St. Peter*, tome x, pp. 151, 152. 1866.

READE, T. M. On the Eozoon canadense. <*Nature*, vol. iii, pp. 146, 147. 1870.

READE, T. M. On the Eozoon canadense. <*Nature*, vol. iii, pp. 267, 367, 368. 1871.

ROWNEY, T. H. On the so-called "Eozoonal" Rock. <*Quart. Journ. Geol. Soc. Lond.*, vol. xxv, pp. 115-118. 1869.

SCHULTZE, M. S. Eozoon canadense. <*Kolner Zeitung*, aug. 14, Cologne. 1873.
> Not seen.

SCHULTZE, M. S. Eozoon canadense. <Ann., and Mag. Nat. Hist., ser. 4, vol. xiii, pp. 324-326. 1874.

SANFORD, (MR.) Announces Eozoon in Connemara Marble of the Binabola Mountains, Ireland. <*Geological Maga*. Reannounced in "Reader," Feb. 25th, 1865.
> Not seen.

THOMSON, W. Palæozoic crinoids. <*Nature*, vol. iv, p. 72. 1871.
    Remark on Eozoon.

VILANOVA Y PEIRA, JUAN. Estructura de las rocas serpentinosas y el Eozoon canadense. *Soc. Espan. Hist. Nat.*, vol. iii, parts 2 and 3. 1874.
    Concludes that Eozoon canadense is not the remains of an organism. (Nicholson in White and Nicholson's Bib. p. 130.)

WHITNEY, J. D., and M. E. Wadsworth. Remarks on the Eozoon from the Azoic System and its subdivisions. <*Bull. Mus. Comparative Zool.*, vol. vii, pp. 528-538. 1884.

WINCHELL, N. H. The cupriferous series in Minnesota. <*Am. Assc. Adv. Sci.* 1880, p. 425. Reprinted in the ninth annual report of the Minnesota survey.
    A remark on the probable Silurian or Cambrian age of the Eozoon-bearing rocks of Canada, based on the age of the Norian rocks of Minnesota.

WINCHELL, N. H. Geology of Minnesota, vol. i, of the final report, p. 283. 1884.
    Note on Eozoon.

PART II.

NORTH AND SOUTH AMERICA, INCLUDING BERMUDA, LEEWARD AND WINDWARD ISLANDS.

NORTH AND SOUTH AMERICA.

ANON. The Nummulites of North America. <*Amer. M. Micro. Journ.*, vol. iv, pp. 1-2. 1883.

ARNOLD, J. W. S. Microscopical Examination of Specimens of Deep Sea Soundings, taken during a cruise of the "Nautical School-ship Mercury," 1871-2. <*Rept. Comm. of Pub. Charities and Corrections*, pp. 13-16. 1872.

AGASSIZ, L. Report upon deep-sea dredgings in the Gulf Stream during the Third Cruise of the United States Steamer Bibb, addressed to Prof. Benjamin Pierce, Superintendent United States Coast Survey. <*U. S. Coast Survey* Report, 1869, pp. 208-219. 1872.

AGASSIZ, A. On the Explorations in the Vicinity of the Tortugas, during March and April, 1881. (Pelagic Fauna of the Gulf Stream). <*Bull. Mus. Comp. Zool.*, vol. ix, pp. 145-149, 1881; also *Nature*, vol xxiv, p. 592. 1881.

BAILEY, J. W. Fossil Foraminifera in the Green Sand of New Jersey. <*Amer. Journ. Sci.*, vol. xli, pp. 213, 214. 1841.

BAILEY, J. W. American Polythalmia from the Upper-Mississippi; and also from the cretaceous formation on the Upper Missouri. <*Amer. Journ. Sci.*, vol. xli, pp. 400, 401; 4 wood cuts. 1841.

BAILEY, J. W. On fossil *Foraminifera* in the calcareous marl from the cretaceous formation on the Upper Missouri, and on silicified wood found near Fredericksburg, Va. <*Proc. Acad. Nat. Sci., Phila.*, vol. i, p. 75. 1843.

BAILEY, J. W. On some New Localities for Infusoria, Fossil and Recent. <*Amer. Jour. Sci.*, vol. xlviii, pp. 321-343. 1845.

BAILEY, J. W. On a Process for Detecting the Remains of Infusoria, etc., in Sedimentary Deposits. <*Proc. Amer. Assoc. Adav. Sci.*, 1849, p. 409. 1850.

BAILEY, J. W. Microscopical examination of Soundings made by the U. S. Coast Survey off the Atlantic Coast of the U. S. <*Smith Contrib. to Knowl.*, vol. ii, p. 15; 1 plate. 1851.

BAILEY, J. W. Observations on a newly discovered animalcule (*Pamphagus*). <*Amer. Journ. Sci.*, vol. xv, 2d ser., pp. 341-347, 1853; and *Quart. Journ. Micro. Sci*, vol. i, pp. 295-299. 1853.

BAILEY, J. W. Examination of some Deep Soundings from the Atlantic Ocean. <*Amer. Jour. Sci.*, vol. xvii, 2d ser., pp. 176-178. 1854.

BAILEY, J. B. On the Origin of Green Sand, and its formation in the Oceans of the present Epoch. <*Proc. Bos. Soc. Nat. Hist.*, vol. v, pp. 364-368, 1856; also in *Amer. Jour. Sci.*, vol. xxii, pp. 280-284. 1856.

BAILEY, J. W. On the Origin of Green Sand and its formation in the Oceans of the present Epoch. <*Quart. Journ. Micr. Sci.*, vol. v, p. 83. 1857.

BAILEY, L. W. Notes on New Species of Microscopical Organisms, chiefly from the Para River, South America. <*Jour. Bos. Sci. Nat. Hist.*, vol. vii, pp. 327-351. 1863.

BARNARD, W. S. Protozoan Studies. <*Proc. Amer. Assoc. Adv. Sci.*, 1871, vol. xxiv, pp. 240-242. (Abstract.) 1872.
> Echinopyxis, Englypha.

BILLINGS, E. Notes on some of the more remarkable genera of Silurian and Devonian fossils. <*Canad. Nat.*, new ser., vol. ii, pp. 184-189, with 14 wood cuts; and pp. 405-409, with 3 wood cuts. 1857.
> Discusses the structure and affinities of Receptaculites, Pascedlas, and Beatricea. (Nicholson in White and Nicholson's Bib. p. 79.)

BILLINGS, E. Palæozoic Fossils, vol. i, 1861-1865. <*Geol. Survey of Canada.* "New Species of Lower Silurian Fossils."

BLAKE, W. P. Notice of Remarkable Strata containing the remains of Infusoria and Polythalamia in the Tertiary Formation of Monterey, California. <*Proc. Acad. Nat. Sci. Phila.*, vol. viii, pp. 328-351. 1856.

BORNEMANN, J. G., in Erman's—Ueber einige bisher nicht beachtete Tertiar-Gesteine aus der Umgegend von Rio de Janeiro. <*Erman's Archiv. v. wissensch, Kunde v. Russland*, vol. xiv, pp. 143-161, pl. iv. 1854.

BRADY, H. B. A monograph of Carboniferous and Permian Foraminifera (the genus Fusulina excepted). <*Palæontographical Society*, 1876, pp. 1-166, plates i-xii.
> This work is necessarily principally concerned with British forms, but not exclusively so. At page 47 is a summary of geological localities in North America which have yielded Carboniferous or Permian *Foraminifera*. The following forms are described from the Carboniferous Rocks of North America: *Valulina palæotrochus*, Eheb., *V. decoerrens*. *V. plicata* Brady. *V. bulloides*, n. sp., *V. rudis*, n. sp., *Nodosinella priscilla*, Dawson. *Calcarina ambigua*. n. sp, and *Endothyra bowmani*, Phill. The last is shown to be the subsequently described *Rotalia baileyi*, Hall, from the Spergen Hill Limestone of Indiana. (Nicholson in White and Nicholson's Bib. p. 86.

BROADHEAD, G. C., in Raphael Pumpelly's Preliminary Report on the Iron Ores and Coal Fields. *Geol. Survey of Missouri.* 1873.
> Fusulina and Receptaculites.

BROADHEAD, G. C. Carboniferous Rocks of Kansas. *Trans. Acad. Sci., St. Louis*, vol. iv, pp. 451-492. 1884.
> Fusulina cylindrica.

CONRAD, T. A. Descriptions of new species of Organic Remains from the Upper Eocene Limestone of Tampa Bay. *Amer. Journ. Sci.*, vol, ii, 2d ser., pp. 399, 400) 9 wood cuts. 1846.
> Describes *N. floridanus, Cristellaria rotella*.

CONRAD, T. A. Observations on the Eocene formation, and descriptions of one hundred and five new fossils of that period, from the vicinity of Vicksburg, Mississippi; with an appendix. *Proc. Acad. Nat. Sci., Phila.*, vol. iii, pp. 280-299. 1848.
> Remarks, among the characteristic fossils, *Nummulites Mantelli N Floridana. Cristellaria rotella.*

CONRAD, T. A. Report of the United States and Mexican Boundary Survey, vol. i, pt. ii. Descriptions of Cretaceous Fossils. 1857.

CONRAD, T. A. Catalogue of the Eocene Annulata, Foraminifera, Echinodermata and Cirripedia of the United States. <*Proc. Acad. Nat. Sci., Phila.*,, vol. xvii, pp. 73-75. 1865.

COUPER, J. H. A letter read by Dr. A. A. Gould, dated at Bainbridge, on the Chatahoochee River, Georgia, March 15, 1845. <*Proc. Bos. Soc. Nat. Hist.*, vol. ii, pp. 123-124. 1848.

 Nummulites, probably *N. Mantelli*.

CRAVEN, AND MAFFIT. Recent Discovery of a Deep-sea Bank on the Eastern Side of the Gulf Stream of the Coast of South Carolina, Georgia and Florida. <*Proc. Amer. Assoc. Adv. Sci.*, vol. vii, pp. 167-171. 1856.

CRISP, F. On Mr. W. B. Thomas' slides of sand obtained by washing clay from the boulder-drift of Meeker county, Minn., U. S. A. <*Journ. R. Micro. Soc.*, ser. ii, vol. iv, p. 504. 1884.

CREDNER, H. Die Kreide von New Jersey. <*Zeitschr. d. deutsch geol. Gesell*, bd. xxii, pp. 191-252. 1870.

CROSBY, W. O. On a Possible Origin of Petrosiliceous Rocks. <*Proc. Bos. Soc. Nat. Hist.*, vol. xx, pp, 160-169. 1879.

CUNNINGHAM, K. M. Cleaning Foraminifera. <*Amer. M. Micro. Journ.* vol. i, p. 88. 1880.

CUNNINGHAM, R. O. Notes on the Natural History of the Strait of Magellan; pp. 28-32. 1871.

DANA, J. D. Descriptions of fossils. <*Appendix to vol. x, Wilkes's U. S. Expl. Exped.*, (Foraminifera from Oregon,) p. 729, pl. 21 of atlas. 1849.

DANA, J. D. Origin of Coral Reefs and Islands. <*Amer. Journ. Sci.*, 3d ser., vol. xxx, pp. 89-105, 169-191. 1885.

DAWSON, G. M. On Foraminifera from Gulf and River St. Lawrence. <*Canad. Nat.*, N. S., vol. v, pp. 172-   wood cuts. 1870.

DAWSON, G. M. Note on the occurrence of Foraminifera, Coccoliths, etc., in the Cretaceous Rocks of Manitoba. <*Canad. Nat.*, new ser., vol. vii, pp. 252-357. 1874.

 The author examined the Cretaceous Rocks of Pembina, some of which resembled the "chalk" of Nebraska in appearance and texture. The earthy base of this deposit consisted principally of *Foraminifera*, coccoliths, and allied organisms. The author describes and figures *Textularia globulosa*, T. pygmæa, *Discorbina globularis*, *Planorbulina ariminensis*, and forms of coccoliths and Rhabdoliths. (Nicholson in White and Nicholson's Bib. p. 91.)

DAWSON, G. M. Report on the geology and resources of the region in the vicinity of the Forty-ninth Parallel, from the Lake of the Woods to the Rocky Mountains; with lists of plants and animals collected, and notes on the fossils. Pages 379, with 18 plates and 3 maps. 1875.

 There are notes on the fossils collected (mostly plants and vertebrates), and amongst these may be mentioned the microscopic organisms (Foraminifera, etc.,) detected by the author in the Cretaceous Rocks of the Pembina escarpment and other localities. (Nicholson in White and Nicholson's Bib. p. 91.)

Dawson, G. M. On a new Species of Loftusia from British Columbia. *Qurt. Journ. Geol. Soc., Lond.*, xxxv, p. 69, pl. vi. 1879.

Dawson, G. M. Bowlder Clays; their Microscopic Structure and the Various Organisms Found in Them. *The Times*, Chicago, June 13, 1885.

Dawson, G. M. Boulder-clays. On the Microscopic Structure of Boulder-clays and the Organisms contained in them. *Bull. Chicago Acad. Sci.*, vol. i, pp. 59-69; 3 wood cuts. 1885. Also in the *Thirteenth Ann. Rept. Geol. and Nat. Hist., Sur. Minn.*, 1884, pp. 150-163; 3 wood cuts. 1885.

Dawson, J. W. Additional Notes on the Post-Pliocene Deposits of the St. Lawrence Valley. <*Canad. Nat.*, vol. iv, pp. 23-39; wood cuts. 1859.
  The author describes and figures 8 species of the Foraminifera.

Dawson, J. W. Notice of Tertiary Fossils from Labrador, Maine, etc., and Remarks on the Climate of Canada in Newer Pliocene or Pleistocene Period. <*Canad. Nat.*, vol. v, pp. 188-200, wood cuts. 1860. Foraminifera.
  The only new species mentioned is *Nonionina labradorica*.

Dawson, J. W. Notes on Post-pliocene Deposits at Riviere Du Loupe and Tadaissac. *Canad. Nat.*, N. S., vol. ii, pp. 81-87. 1865.

Dawson, J. W. On Foraminifera from the Gulf and River St. Lawrence. <*Amer. Journ. Sci.*, vol. i, ser. 3, pp. 204-210; 10 wood cuts. 1871.

Dawson, J. W. On Foraminifera from the Gulf and River St. Lawrence. <*Ann. and Mag. Nat. Hist.*, ser. 4, vol. vii, pp. 83-89. 1871.

Dawson, J. W. On Some New Specimens of Fossil Protozoa from Canada. *Proc. Amer. Assoc. Adv. Sci.*, 1875, vol. xxiv, pp. 100-105; wood cuts. 1872.

Dawson, J. W. Notes on the Post-pliocene Geology of Canada. *Canad. Nat.*, N. S., vol. vi, pp. 19, 166, 241, 369, pl. iii. 1872.

Dawson, J. W. On some New Specimens of Fossil Protozoa from Canada. *Proc. Amer. Assoc. Adv. Sci.*, vol. xxiv, pp. 100-106. 1876.

Dawson, J. W. Palæontological Notes.—II. Saccammina? (Calcisphæra) Eriana. *Canad. Nat.*, vol. x, No. 1. 1881.

D'Orbigny, A. Voyage dans l'Amerique Meridinale pendant les Annees, 1826-1833. Paris, 1834-'43, vol. v, partie 5. *Foraminiferas*, fol. 9 pl's. 1839.

D'Orbigny, A. D. Die Foraminiferen Amerikas und der Canarischen Inseln. (Muller Archiv.) 80 Berlin. 1840.

Ehrenberg, C. G. Verbreitung und Einfluss der Mikroscopischen Lebens in Sud und Nord Nord Amerika. *Abhan. Kongl. Akad. Wiss.* Berlin, (1841), pp. 291, 438; 4 plates. 1841.

Ehrenberg, C. G. Verbreitung des Mikroskopischen Lebens als Felsmassen im Centralen Nord-Amerika und im Westlichen Asien. *Berichte d. k. preuss. Ak. Wiss.*, 1842, pp. 187, 188. 1842.

Ehrenberg, C. G. Ueber das mikroscopische Leben in Texas. *Berichte d. Kongl. preuss. Akad. Wiss.*, (1849), pp 87-91. 1849.

EHRENBERG, C. G. Verbreitung und Einfluss des mikroskopischen Lebens in Sud und Nord Amerika. *Abhand. d. Akad. d. Wiss. zu Berlin* (1841), pp. 291-445; 4 plates. 1843.

EHRENBERG, C. G. Report on the species of Infusoria contained in specimens of the sediment of the Mississippi river. *Astronom. Obser. Nat. Obser. Wash.*, vol. III, appendix B. Observations on the Mississippi River at Memphis, Tenn., pp. 26-32. 1853.

EHRENBERG, C. G. Die weitere Entwickelung Kenntniss des Grundsandes als grune Polythalamien-Steinkerne, ueber braunrothe und corall-rothe Steinkerne der Polythalamien-Krede in Nord-Amerika, und ueber den Meeresgrund aus 12,900. Fuss Tiefe. [The further development of the discovery that the green sand is composed of green casts of polythalmia, also on the brownish-red or bright-red casts of polythalmia in chalk of North America, and on the sea bottom at depths of 12,900.] *Monatsbericht d. kk. Akad. d. Wiss. Berlin*, 1855, pp. 172-178.

> The chief point in this paper is that the brownish or reddish "chalk" of Alabama owes its color to numerous shells of *Foraminifera* filled with a similarly colored silicate of iron. Nicholson in White and Nicholson's Bib. p. 98.)

EHRENBERG, C. G. Erlauterungen ueber den Grunsand im Zeuglodon-Kalke Alabam's in Nord-Amerika. [Investigations into the green sand of the Zeuglodon limestone of North America.] *Monatsbericht d. k. k. Akad. d. Wiss. Berlin*, 1855, pp. 86-89.

> The author shows that the grains of green sand interspersed in the Zeuglodon-limestone of Alabama are really of the nature of casts of the shells of Polythalamons *Foraminifera*. At least thirty different forms were recognized by the author. (Nicholson in White and Nicholson's Bib. p. 98.)

EHRENBERG, C. G. Beitrag zur Uebersicht der Elemente des tiefen Meeresgrundes im Mexicanischen Golfstrome bei Florida. *Monatsber d. k. pr. Akad. d. Wiss. Berlin* (1861), pp. 222-240; tables. 1862.

FABRICIUS, O. Fauna Groenlandiæ, systematice sistens animalia Groenlandiæ occidentalis hactenus indagata, etc. Hafniæ et Lipsiæ. 1780.

GABB, W. M. Descriptions of new species of American Tertiary and Cretaceous Fossils. *Journ Acad. Sci., Phila.* n. s., vol. iv, pp. 375-406, pl. lxix; 1860.

GABB, W. M. Catalogue of the Invertebrate Fossils of the Cretaceous formation of the United States, with references. *Proc. Acad. Nat. Sci., Phila.*, 1859, 20 pages. 1860.

GABB, W. M. Description of a Collection of Fossils made by Dr. Antonio Raimondi in Peru. *Journ. Acad. Nat. Sci., Phila.*, vol. viii, pp. 263-336. 1877.

GALEOTTI, H. G. Sur le calcaire Cretace des environs de Jalapa au Mexique. <*Bull. de la Societe Geol. de France*, vol. x. 8 vo. Paris. 1839.

GEINITZ, H. B. Carbonformation und Dyas in Nebraska. <*Acta Academia Leop. Carol.*, vol. xxxiii, pp. 1-91; 5 plates. 1866.

> Fusulina depressa, Fischer. F. cylindrica, Fischer.

HALL, J. Description of new species of Fossils from the Carboniferous Limestones of Indiana and Illinois. *Trans. Albany Inst.*, iv, pp. 2-36. 1856.
    Rotalia Baileyi, p. 34.

HALL, J. Observations upon the Cretaceous Strata of the United States with reference to the Relative Position of Fossils Collected by the Boundary Commission. *Amer. Journ. Sci.*, vol. xxiv, 2d ser., pp. 72-86. 1857.

HALL, J. Notice of some New Species of Fossils from a Locality of the Niagara Group, in Indiana, with a list of Identified species from the same place. *Trans. Albany Inst.*, iv, pp. 195-228. 1862.
    Receptaculites subturbinatus (Hall) p. 224.

HAMLIN, F. M. The Preparation and Mounting of Foraminifera, with Description of a New Slide for Opaque Objects. *Proc. Amer. Soc. Mic.*, sixth meeting, 1883, pp. 65-68. 1883.

HARPER, L. Preliminary Report on the Geology and Agriculture of the State of Mississippi, pp. 348; tables I-VII. 1857.

HARVEY, W. H., and J. W. BAILEY. New Species of Diatomaceæ, collected by the United States Exploring Expedition, under the command of Captain Wilkes, U. S. N. Appendix. (Lagena Williamsoni.) *Proc. Acad. Nat. Sci., Phila.*, vol. vi, p. 430. 1853.

HAYDEN, (DR.) F. V. Geological Report of the Yellowstone and Missouri River's Foraminifera, p. 123. 1860.

HAYDEN, (DR.) F. V. Final Report of the United States Geological Survey of Nebraska and Portions of the Adjacent Territories, p. 140, pl. ii, v. 1872.

HEILPRIN, A. Notes on the Tertiary Geology of the Southern United States. *Proc. Acad. Nat. Sci., Phila.*, 1881, pp. 151-159. 1881.

HEILPRIN, A. On the Occurrence of Nummulitic Deposits in Florida, and the Association of Nummulites with Fresh-water Fauna. *Proc. Acad. Nat. Sci., Phila.*, pp. 189-194. 1882.

HEILPRIN, A. Notes on some New Foraminifera from the Nummulitic Formation of Florida. *Proc. Acad. Nat. Sci., Phila.*, 1884, pp. 321-322. 1884.

HEILPRIN, A. The Tertiary Geology of the Eastern and Southern United States. *Journ. Acad. Nat. Sci., Phila.*, 2d ser., vol. iv, pp. 115-154. 1884.

HILGARD, E. W. Remarks on the new division of the Eocene, or Shell Bluff Group, proposed by Mr. Conrad. *Amer. Jour. Sci*, vol. xlii, 2d ser., pp. 68-70. 1866.

HILGARD, E. W. On the Tertiary Formations of Mississippi and Alabama. *Amer. Journ. Sci.*, vol. xliii, 2d ser., pp. 29-41. 1867.

HILGARD, E. W. On the Geology of the Delta, and the Mud-lumps of the Passes of the Mississippi. *Am. Jour. Sci.*, vol. i, 3d ser., pp. 425-437. 1871.

HITCHCOCK, C. H. Notes on the Geology of Maine. *Proc. Portland Soc. Nat. Hist.*, vol. i, pp. 72-86. 1862.

HITCHCOCK, R. Synopsis of the Fresh-water Rhizopods, 8 vo. 2381;

HITCHCOCK, R. The Cause of Variation. *Ann. and Mag. Nat. Hist.*, ser 5, vol. xiv, pp. 93-97. 1884.

HOPKINS, F. V. Report on the Microscopic examination of the specimens. <*Reclamation of the Alluvial Basin of the Mississippi River.* Appendix No. 2. 1878.

HOPKINS, F. V. List of Microscopic Organisms, with two plates. *Reclamation of the Alluvial Basin of the Mississippi.* Appendix No. 4. 1878.

HONEYMAN D. Chebucto Nullipores, with Attaches. <*Proc. Trans. Nova Scotia Inst. Nat. Sci.*, vol. vi, 1882-83., pp. 8-12. 1883.

JAMES, F. L. Separation of the Sand from Diatoms and Foraminifera. <*The Micro. Bull. Sci. News*, vol. ii, p. 43. 1885.
See also *National Druggist.* p. 60, vol. v.

JAMES, J. F. Remarks on the Genera Lepidoiites, Anomaloides, Ischadites and Receptaculites, from the Cincinnati Group. <*Journ. Cin. Soc. Nat. Hist.*, vol. viii, pp. 163-166. 1885.

JAMES, T. R., and W. R. PARKER. On the Foraminifera of the Family Rotalinæ (Carpenter) found in the Cretaceous Formations; with Notes on their Tertiary and Recent Representatives. <*Quart. Journ. Geol. Soc. Lond.*, vol. xxviii, pp. 103-131. 1872.

The American forms treated of in this communication are the Cretaceous Rotalines described by Ehrenberg, from the Missouri and Mississippi (*Mikrogeologie*), and those described by Reuss from the Green sand of New Jersey. (*See Reuss.*) (Nicholson in White and Nicholson's Bib. p. 110.)

JOHNSON, DR. H. A., and B. W. THOMAS. Report of the Committee on the Microscopic Organisms in the Bowlder Clays of Chicago and vicinity. <*Bull. Chic Acad. Sci.*, vol. i, No. 4, pp. 35-40. 1884.

KARRER, F., L. F. Pourtales. Der Boden des Golfstroms und der Atlantischen Kuste Nord-Amerika's (Petermann's Mittherlungen 16 Bd. 1870, XI.) <*K. k. Geol. Reich.* Ver. 1870, pp. 329-331. 1870.

LEA, ISAAC. Contributions to Geology, 227 pp.; 6 plates. 1833. Miliola Maryladica, p. 215, pl. vi.

LEIDY, J. Remarks on some Marine Rhizopoda. <*Proc. Acad. Nat. Sci., Phila.*, 1875, p. 73. 1875.

LEIDY, J. Foraminiferous Shells of our Coast. <*Proc. Acad. Nat. Sci., Phila.*, 1878, p. 336 1878.

LEIDY, J. Foraminifera of the Coast of New Jersey. <*Proc. Acad. Nat. Sci., Phila.*, 1878, p. 292. 1878.

LEIDY, J. Fresh-water Rhizopods of North America, 4to 48 plates. 1879.

LEIDY, J. Foraminifera in the Drift of Minnesota. <*Proc. Acad. Nat. Sci., Phila.*, 1884, pp. 22, 23. 1884.

LYELL, C. Notes on the Cretaceous Strata of New Jersey and other Parts of the United States bordering the Atlantic. *Quart. Jour. Geol. Soc.*, vol. i, pp. 55-60, 1845.

LYELL, C. Notice of the Foraminifera of New Jersey. <*Quart. Journ. Geol. Soc., Lond.*, vol. i, p. 64. 1845.

LYELL, C. On the Newer Deposits of the Southern States of North America. <*Quart. Jour. Geol. Soc., Lond.*, vol. ii, pp. 405–410. 1846.
    Nummulites Mantelli.

LYELL, C. Om the relative Age and Position of the so-called Nummulite Limestone of Alabama. <*Amer. Journ. Sci.*, vol. iv, 2d ser., pp. 186-191. 1847.

LYELL, C. On the relative age and position of the so-called Nummulite Limestone of Alabama. <*Quart. Journ. Geol. Soc., Lond.*, vol. iv, pp. 10–16. 1848.
    Numerous fossils are alluded to as occurring in the strata in question, and the memoir contains notes from Edward Forbes and Alcide D'Orbigny as to the zoological position of *Orbitoides* (*Nummulites*) *mantelli*. Nicholson in White and Nicholson's Bib. p. 114.)

MAURY, M. F. The Physical Geography of the Sea, pp. 274; 12 plates. 1855.
    Containing much interesting matter, on the sea bottom.

MEEK, F. B., and Dr. F. V. HAYDEN. Remarks on the Lower Cretaceous Bed of Kansas and Nebraska, together with descriptions of some new species of Carboniferous Fossils from the Valley of Kansas River. <*Proc. Acad. Nat. Sci., Phila.*, pp. 256–266. 1858.

MEEK, F. B., and F. V. HAYDEN. Geological Explorations in Kansas Territory. *Proc. Acad. Nat. Sci., Phila.*, 1859, pp. 8–30. 1860.

MEEK, F. B. In A. H. Worthen's *Geological Survey of Illinois*, vol. v. Palæontology, p. 560; pl. xxiv. 1873.

MEEK, F. B. and A. H. WORTHEN. Descriptions of new Species and Genera of Fossils from the Palæozoic rocks of the Western States. *Proc. Acad. Nat. Sci., Phila.*, 1870, pp. 22–56. 1870.
    Receptaculites pp. 22, 23.

MEYER, O. The Genealogy and the Age of the species in the Southern Old-tertiary. *Amer. Jour. Sci.*, 3rd ser., vol. xxiv, pt. I, pp. 457–468. II, vol. xxx, pp. 60–72. III, vol. xxx, pp. 1-16. 1885.

MORTON, S. G. Supplement to the "Synopsis of the Organic Remains of the Ferruginous Sand Formation of the United States." *Amer. Journ. Sci.*, vol. xxii, pp. 288-294; 2 plates. 1833.
    Figure and description of *Nummulites Mantelli*.

MORTON, S. G. Synopsis of the Organic Remains of the Cretaceous Group. 8 vo. Philadelphia, 1834.

MORTON, S. G. Description of some new Species of Organic Remains of the Cretaceous group of the United States, with a tabular view of the Fossils hitherto discovered in this formation. *Jour. Acad. Nat. Sci.*, vol. viii, pp. 207–227. 1842.
    Planularia comcata.

MURRAY, J. Reports on the Results of Dredging, under the Supervision of Alexander Agassiz, in the Gulf of Mexico (1877-'78), in the Carribean

1878-'79), and along the Atlantic Coast of the United States, during the Summer of 1880, by the U. S. Coast Survey Steamer "Blake" Lieutenant-Commander C. D. Sigsbee, U. S. N , and Commander J. R. Bartlett, U. S. N., Commanding. XXVII. Report on the Specimens of Bottom Deposits. <*Bull. Mus. Comp. Zool.*, vol. xii, No. 2, pp. 1-61. 1885.

OWEN, D. D. Geological Survey of Wisconsin, Iowa, and Minnesota; pp. 586, 587, pl. ii B. 1852.

(Foraminifera.) Selenoides. (N. G.?)

PACKARD, A. S., (Jun.) A list of Animals dredged near Caribou Island, South Labrador, during July and August, 1860, with a list of the Invertebrata collected at Anticosti and Mingan Islands by A. E. Verrill, etc. <*Canad. Nat.*, vol. viii, pp. 400-429. 1863.

PACKARD, JR., A. S. Life History of the Protozoa. <*Amer. Nat.*, vol. viii, pp. 728-748. 10 wood cuts. 1874.

POURTALES, L. F., DE. On the order of Succession of Parts in Foraminifera, Communicated by Prof. Agassiz. *Proc. Amer. Assoc. Adv. Sci.*, vol. iii, p. 89. 1850.

POURTALES, L. F., DE. On the Distribution of the Foraminifera on the Coast of New Jersey, as shown by the off-shore soundings of the Coast Survey, Communicated by Prof. A. D. Bache. <*Proc. Amer. Assoc. Adv. Sci.*, 1850, pp. 84-88. 1850.

POURTALES, L. F., DE. Notes on the Specimens of the Bottom of the Ocean brought up in recent Explorations of the Gulf Stream, in Connection with the Coast Survey. *Proc. Amer. Assoc. Adv. Sci.*, 1853, vol. vii, pp. 181-184. 1856.

POURTALES, L. F. On the genera Orbulina and Globigerina of D'Orbigny. <*Amer. Journ. Sci.*, vol. xxvi, 2d ser., p. 96. 1858.

POURTALES, L. F., DE. Contributions to the Fauna of the Gulf Stream at great depths. <*Bull. Mus. Comp. Zool.*, No. 6, 1867, p. 103. 1863-69.

POURTALES, L. F. Der Boden des Gulfstromes und der atlantischen Kuste Nord—America's. <*Petermann's Geogr. Mittheil.*, vol xvi, pp. 393-398. 1870.

POURTALES, L. F. The Gulf Stream,—Characteristics of the Atlantic Sea-bottom off the coast of the United States. *U. S. Coast Survey Report*, 1869, pp. 220-225, 1872.

REUSS, A. E. Die Foraminifera des Senonischen Grunsandes von New Jersey. Palæontologische Beitrage. <*Sitzungsb. Math-Naturn. cl. Kais. Akad. Wiss. Wien*, vol. xliv, pp. 334-340, pl. vii, fig. 6, and pl. vii, fig. 1. 1861.

Describes and figures *Rotalia mortoni* and *Truncatulina Dekayi*. (Nicholson in White and Nicholson's Bib. p. 124.)

ROEMER, F. Die Kreidebildungen von Texas. 4to, 11 plates. Washington, 1852.

RYDER, J. A. The Protozoa and Protophytes Considered as the Primary or Indirect Source of the Food of Fishes. <*Bull. U. S. Fish. Comm.*, vol. i, pp. 236-251. 1881.

SALTER, J. W. Fossils from the base of the Trenton Limestone. <*Figures and Descriptions of Canadian Organic Remains.* Decade I, Montreal, 1859, pp. 47, pls. i-x.

> The author deals with the affinities and structure of the genus *Receptaculites*, referring the fossils of this group to the *Foraminifera*, and placing them in the neighborhood of *orbitalites*. Two new species are described, one, R. occidentalis, from the Trenton Limestone, and the other, R. australis, introduced for comparison for the Silurian rocks of New South Wales. (Nicholson in White and Nicholson's Bib. p. 127.)

SCHLUMBERGER, C. Remarks upon a species of Cristellaria. *Journ. Cinn. Soc. Nat. Hist.*, vol. v, p. 119, plate 5, figs. 2, 2a. 1882.

SHUMARD, B. F. Notice of New Fossils from the Permian Strata of New Mexico and Texas, collected by Dr. George G. Shumard, Geologist of the United States Government Expedition for obtaining water by means of Artesian Wells along the 32d parallel, under the direction of Capt. John Pope, U. S. Corps, Top. Eng. < *Trans. Acad. Sci. St. Louis*, vol. i, p. 297. 1858.

SMITH, E. A. On the Geology of Florida. <*Amer. Journ. Sci.*, 3d ser., vol. xxi, pp. 292-309. 1881.

SMITH, E. A. Remarks on a paper of D'r Otto Meyer on "Species in the Southern Old-Tertiary." <*Amer. Journ. Sci.*, 3d ser., vol. xxx, pp. 270-275. 1885.

SPENCER, J. W. Stromatoporidæ of the Upper Silurian System. <*Bull. Mus. Univ. S. Missouri.* Part II, vol. i, No. i, pp. 43-53, pl. vi. 1884. Also *St. L. Acad. Sci.*, vol. iv, No. 4.

VERNEUIL, E. de. On the *Fusulina* in the coal formation of Ohio. *Amer. Journ Sci.*, vol. ii, 2d series, p. 293. 1846.

VERNEUIL, (de) P. E. Note sur le parallelisme des roches des depots paleozoiques de l'Amerique Septentrionale avec ceux de l'Europe. <*Bull. Soc. Geol. de France*, ser. 2, vol. iv, p. 682. 1847.

VERRILL, A. E. Materials of Sea Bottoms. Their nature and origin in the region of the Gulf Stream. <*N. Y. Sunday Times.* Feby. 1883.

VORCE, C. M. Cleaning Foraminifera. *Amer. Month. Micr. Journ.*, vol. i, p. 24. 1880.

WALLICH, —. Critical observations on Prof. Leidy's "Freshwater Rhizopods of North America," and classification of the Rhizopods in general. <*Ann., and Mag. Nat. Hist.*, 5 ser., vol. xvi, pp. 317-334, 453-473. 1885.

WHITE, C. A. Note on Endothyra ornata. *Proc. U. S. National Mus.* 1879, p. 291. 1879.

WHITE, C. A., and ST. JOHN, O. H. Description of New Sub-carboniferous and Coal-Measure Fossils, collected upon the Geological Survey of Iowa. *Trans. Chicago Acad. Sci.*, vol. i, pp. 115-127. 1867.

WHITEAVES, J. F. On some Results obtained by Dredging in Gaspe and off Murry Bay. <*Canad. Nat.*, n. s. vol. iv, p. 270. 1869.

WHITEAVES, J. F. Notes on a Deep-sea Dredging-Expedition round the Island of Anticosti, in the Gulf of St. Lawrence. *Ann., and Mag. Nat. Hist.*, ser. 4, vol. x, pp. 341-354. 1872.

WHITEAVES, J. F. Report on a Deep sea Dredging Expedition to the Gulf of St. Lawrence. <*Appendix K. Fourth Ann. Rept. Dept. Marine and Fisheries*, pp. 90-101. 1872.

WHITEAVES, J. F. Notes of a Deep-sea Dredging Expedition round the Island of Anticosti, in the Gulf of St. Lawrence. *Brit. Assoc. Advan. Sci.* 1872. Pp. 143-145. 1873.

WHITEAVES, J. F. Report on Deep Sea Dredging Operations in the Gulf of St. Lawrence, with notes on the present conditions of the Marine Fisheries and Oyster beds of part of that Region. <*Appendix U. Ann. Rept. Dept. Marine and Fisheries*, pp. 178-204. 1874.

WHITEAVES, J. F. On recent Deep-Sea Dredging Operations in the Gulf of St. Lawrence. <*Amer. Journ. Sci.*, vol. vii, 3d ser., pp. 210-219. 1874.

WHITFIELD, R. P. On the Fauna of the Lower Carboniferous limestones of Spergen Hill, Ind., with a revision of the descriptions of its Fossils hitherto published, and illustrations of the species from the original type series. <*Bull. Amer. Mus. Nat. Hist*, vol. i, No. 3, pp. 39-97, pls. vi-ix. 1882.

    Endothyra Baleyi, Hall's sp. (?) Figs. 34-36, p. 42.

WOODWARD, A., and B. W. THOMAS. On the Foraminifera of the Boulder-Clay, taken from a well-shaft 22 feet deep, Meeker County, Central Minnesota. <*Thirteenth Ann. Rept. Geol., and Nat. Hist. Surv. Minn.*, 1884. pp. 164-177; pls. iii, iv. 1885.

BERMUDA.

WOODWARD, A. Foraminifera of Bermuda. *Journ. N. Y. Micro. Soc.*, vol. i, pp. 147-151. 1885.

LEEWARD AND WINDWARD ISLANDS.

BURY, MRS. Polycystius; figures of remarkable forms, etc., in the Barbadoes Chalk Deposit. 2nd edition. Edited by M. C. Cooke. 4 to. London. 1862.

BURY, P. S. Polycystius; Remaakable Forms, etc., from the Barbadoes deposit, second edition, edited by M. C. Cooke. 4 to. 1867.

    Only fifty copies produced of this very curious microscopic work.

D'Orbigny, A., In M. Ramon de la Sagra Histoire de l'Ile de Cuba. Foraminiferes, 224 pp., and atlas. 8 vo., folio. Paris, 1839.

D'Orbigny, A. Foraminiferes. In Ramon de la Sagra's Histoire physique, politique et naturelle de l'Ile de Cuba. French edition, 8 vo., 1839; Spanish edition, 1840, fol., 12 plates.

Duncan, P. M. A notice of the Geology of Jamaica, especially with reference to the District of Clarendon; with Descriptions of the Cretaceous, Eocene, and Miocene Corals of the Islands. *Quart. Journ. Geol. Soc. Lond.*, vol. xxi, pp. 1-15, 2 plates. 1865.

Ehrenberg, C. G. Ueber die mikroskopischen kieselschaligen Polycystinen als machtige Gebirgsmasse von Barbados; *Monatsberichte der Konige, Akad, der Wissenschaften zu Berlin.* 1847.

Ehrenberg, C. G. Fortsetzung der mikrogeologischen Studien als Gesammt-Uebersicht der mikroskopischen Palaontologre gleichartig analysister Gebirgsarten der Erde, mit specieller Rucksicht auf den Polycystinen Mergel bei Barbados. *Abhand. d. Akad. d. Wiss. Berlin*, 1875, pp. 1-168, 30 plates. 1875.

Guppy, R. J. L. On the occurrence of Foraminifera in the Tertiary beds of San Fernando, Trinidad. *Trans. Sci. Assoc. Trinidad*, 1863, p. 11, also Geologist, 1864, p. 159.

Guppy, R. J. L. On the Tertiary Mollusca of Jamaica. *Quart. Journ. Geol. Soc. Lond.*, vol. xxii, pp. 281-295, 2 plates. 1866.

Guppy, R. J. L. On Tertiary Brachiopoda from Trinidad. *Quart. Journ. Geol. Soc. Lond.*, vol. xxii, pp. 295, 296. 1866.

Guppy, R. J. L. On the Relations of the Tertiary Formations of West Indies, with a note on a new species of Rannia by Henry Woodward and on the Orbitoides and Nummulina by T. Rupert Jones. *Quart. Journ. Geol. Soc. Lond.*, vol. xxii, p. 570. 1866.

Guppy, R. J. L. On the discovery of Organic Remains in the Caribbean Series of Trinidad. *Quart. Journ. Geol. Soc. Lond.*, vol. xxvi, pp. 413-415. 1870.

Guppy, R. J. L. On the Foraminifera from the Tertiaries of San Fernando, Trinidad. *Proc. Sci. Assoc. Trinidad.*, pp. 13-16, 1872; also *Geol. Mag.*, dec. I, vol. x, pp. 362-363. 1873.

Guppy, R. J. L. On the West Indian Tertiary Fossils. *Geol. Mag.*, 1874, p. 445.

A list of the Foraminifera of the Tertiary Deposits of the West Indies.

Heneken, T. S. On some Tertiary Deposits in San Domingo. *Quart. Journ. Geol. Soc. Lond.*, vol. ix, pp. 115-129. 1853.

Jones, T. R. Note on some Nummulinae and Orbitoides from Jamaica. *Quart. Jour. Geol. Soc. Lond.*, vol. xix, pp. 514, 515. 1863.

Jones, T. R. The Relationship of certain West-Indian and Maltese strata, as shown by some Orbitoides and other Foraminifera. *Geol. Mag.*, dec. I, vol. i, pp. 102-106. 1864.

Jones, T. R. In Guppy's Relations of the Tertiary formations of the West Indies. On the Orbitoides and Nummulinae. <*Quart. Journ. Geol. Soc. Lond.*, vol xxii, pp. 570-593, pl. xxvi. 1866.

Jones, T. R. Note on the Orbitoides and Nummulinæ of the Tertiary Asphaltic Bed, Trinidad. *Quart. Journ. Geol. Soc. Lond.*, vol. xxii, pp. 572, 573. 1866.

Jones, T. R., and W. K. Parker. Notes on some Fossil and Recent Foraminifera collected in Jamaica by the late Lucas Barrett, F. G. S. <*Report. Brit. Assoc.* (Newcastle-on-the-Tyne Meeting) Trans. Sections, p. 80. 1863.

Jones, T. R., and W. K. Parker. Note on some Foraminifera dredged by the late Mr. Lucas Barrett at Jamaica. <*Report Brit. Assoc.* (Newcastle-on-Tyne Meeting.) Trans. Section. p. 105. 1863.

Jones, T. R. and Parker, W. K. Notice sur les Foraminiferes vivants et Fossiles de la Jamaique. Bruxelles, 1876.

Moore, J. C. On some Tertiary Beds in the Island of San Domingo; from Notes by J. S. Heniker. <*Quart. Journ. Geol. Soc. Lond.*, vol. vi, pp. 39-53, 2 plates. 1850.

Moore, J. C. Notes on the Fossil Mollusca and Fish from San Domingo. <*Quart. Journ. Geol. Soc. Lond.*, vol. ix, pp. 129-132. 1853.

Moore, J. C. On some Tertiary Shells from Jamaica. <*Quart. Journ. Geol. Soc. Lond.*, vol. xix, pp. 510-513. 1863.

Schomburgh, (Sir) R. H. The Microscopical siliceous Polycystina of Barbadoes, and their relation to existing animals as described in a lecture by Prof. Ehrenberg of Berlin, delivered before the Royal Acad. of Sci. 1847. <*Ann., and Mag. Nat. Hist.*, ser. I, vol. xx, pp. 115-127. 1847.

Schomburgh, R. H. The History of Barbadoes, 772, pp. 8vo. London, 1848.

Van Broeck, E. Etude sur les Foraminiferes de la Barbade (Antilles) recueillis par L. Agassiz precedee de quelques considerations sur la classification et la nomenclature des Foraminiferes. <*Ann. de la Soc. Belg. de Micro.*, vol. ii, pp. 68-152, 2 plates. 1876.

PART III.

———

ENGLAND, IRELAND, SCOTLAND
AND WALES.

ENGLAND, IRELAND, SCOTLAND AND WALES.

ADAMS, G. Micrographia Illustrata; or the knowledge of the Microscope explained. 4to. London, 1747. Fourth edition, with 72 plates, 8 vo. in 1771.

ADAMS, G. (filius). Essays on the Microscope, containing a description of the most improved microscopes, a general history of insects, and a description of 379 animalculæ, etc. 4to. London, 1787. A second edition in 1798, edited by Frederick Kanmacher. 4to., with folio plates.

ADAMS, J. Descriptions of some minute British Shells.. ⁻Trans. Linnaean Soc., Lond., vol. v, pp. 1-6; pl. i. 1800.

ALCOCK, T. On Natural History Specimens recently received from Connemara. <Proc. Lit., and Phil. Soc., Manches., vol. iv, p. 193; wood cuts. 1865.

ALCOCK, T. Foraminifera of Dogs Bay. Proc. Lit. Philos. Soc., Manchester, vol. v, pp. 99, 100. 1866.

ALCOCK, T. On Foraminifera from a Shell of Halia Priamus. <Proc. Lit. Philos. Soc., Manchester, vol. v, p. 123. 1866.

ALCOCK, T. On Polymorphina tubulos. ⁻Proc. Lit. and Phil. Soc., Manchester, vol. vi, p. 85. 1867.

ALCOCK, T. Questions regarding the Life-History of the Foraminifera, suggested by Examinations of their Dead Shells. ⁻<Mem. Lit. Philos. Soc., Manchester, ser. iii, vol. iii, pp. 175-181; 1 plate. 1868.

ALLMAN, G. J. Note on Polytrema miniaceum. Ann. and Mag. Nat. Hist., ser. 4, vol. v, pp. 372, 373. 1870.

ALLMAN, P. Recent Researches among some of the more simple Sarcode Organisms. ⁻Journ. Lin. Soc., Lond., vol. viii, pp. 261-305; 19 wood cuts, pp. 385-439, 17 wood cuts. 1878.

ANON. On the natural Position and Limits of the group Protozoa. ⁻Nat. Hist. Review, 1861, pp. 34-43.
    A review.

ANON. Localities for Marine Foraminifera. ⁻Journ. R. Micr. Soc., Lond., vol. iii, p. 497. 1880.

ANON. Importance of Foraminifera for the Doctrine of Descent. ⁻Journ. R. Micr. Soc., Lond., vol. iii, p. 975. 1880.

ANON. Orbulina universa. <Journ. Micr. Soc., ser. ii, vol. iv, pp. 579, 580. 1884.

ANSTED, D. T. On the Geology of the Southern Part of Andalusia, between Gibraltar and Almeria. <*Quart. Journ. Geol. Soc.*, *Lond.*, vol. xiv, pp. 130-133. 1858.

ARMSTRONG, J., J. YOUNG, and D. ROBERTSON. Catalogue of the Western-Scottish Fossils. 8 vo. Glasgow. 1876.

BALKWILL, F. P., and JOSEPH WRIGHT. Recent Foraminifera of Dublin and Wicklow. <*Proc. Royal Irish Acad.* 1882.
A Preliminary Report; not seen.

BALKWILL, F. P., and F. W. MILLET. The Foraminifera of Galway, Pt. I. <*Journ. of Microscopy and Nat. Sci.*. vol. iii, pp. 19-28, pls. i-iv. 1884.

BALKWILL, F. P., and J. WRIGHT. Recent Foraminifera of Dublin and Wicklow. <*Proc. R. Irish Acad.*, ser. 2, vol. iii, pp. 545-550. 1882.

BAUERMAN, H. On the Occurrence of Celestine in the Nummulitic Limestone of Egypt. <*Quart. Journ. Geol. Soc. Lond.*, vol. xxv, pp. 40-44. 1869.

BENNIE, J. Note on the Range of Saccammina carteri (Brady) in the Carboniferous Series. <*Geol. Mag.*, n. s. dec. II, vol. iii, p. 47. 1876.

BIGSBY, J. J. Thesaurus Siluricus, 4to, London. 1868.

BIGSBY, J. J. Thesaurus Devonico—Carboniferous, 4to, London. 1878.

BLAKE, J. F. On the Infralias in Yorkshire. With an appendix on some Bivalve Entomostraca, by Prof T. Rupert Jones, F. G. S. <*Quart. Journ. Geol. Soc. Lond.*, vol. xxviii, pp. 132-147. 1872.

BLAKE, J. F. On the Kimmeridge Clay of England. *Quart. Journ. Geol. Soc. Lond.*, vol. xxxi, pp. 196-233. 1875.

BLAKE, J. F. Lower-Silurian Foraminifera. *Geol. Mag.*, n. s., dec. II, vol. iii, p. 134. 1876.

BLAKE, J. F. On Renulina Sorbyana. *Monthly Micr. Journ.*, vol. xv, p. 262, wood cut. 1876.

BOWDICH, T. E. Elements of Conchology, including the Fossil Genera and the Animals, p. 75. Paris, 1822.

BOWERBANK, J. S. On the Anatomy and Physiology of the Spongiadæ. *Philos. Trans.* p. 279. 1858.

BRADY, H. B. Report on the Dredging of the Northumberland and Coast and Dogger Bank, drawn up by Henry T. Mennell. *Brit. Assoc. Advan. Sci.* (Foraminifera), 1862, p. 122. 1863.

BRADY, H. B. Notes on Foraminifera new to the British Fauna. *Report Brit. Assoc.* (Newcastle-on-Tyne Meeting), Trans. Section, p. 100. 1863.

BRADY, H. B. Contributions to Knowledge of the Foraminifera. On the Rhizopodal Fauna of the Shetlands. <*Trans. Linn. Soc. Lond.*, vol. xxiv, p. 463, pl. xlviii. 1863.

BRADY, H. B. Foraminifera;—in report of Deep-sea Dredging on the Coasts of Northumberland and Durham in 1862–1864. <*Nat. Hist. Trans. Northd., and Durham*, vol. i, p. 51. 1863.

BRADY, H. B. On Involutina liassica (Nummulites liassicus, Rupert Jones). *Geol. Mag.*, vol. i, p. 193, pl. ix. 1863.

BRADY, H. B. On the Foraminifera of the Middle and Upper Lias of Summersetshire. *Brit. Assoc. Advan. Sci.* (Bath Meeting), Trans. Section, 1863, p. 50.

BRADY, H. B. A catalogue of the Recent Foraminifera of Northumberland and Durham. <*Nat. Hist. Trans. Northd., and Durham*, vol. i, p. 83, pl. xii. 1865.

BRADY, H. B. Notes on Foraminifera from the Valley-deposits of the Nar., Norfolk. <*Geol. Mag.*, vol. ii, pp. 306, 307. 1865.

BRADY, H. B. On the Rhizopodal Fauna of the Hebrides. <*Brit. Assoc. Advan. Sci.* (Nottingham meeting), 1866, pp. 69, 70. 1866.

BRADY, H. B., in C. W. PEACH's—Further Observations on, and additions to, the list of Fossils found in the Boulder Clay of Caithness. N. B. *Report Brit. Assoc.* (Nottingham meeting), p. 64. 1866. See Peach.

BRADY, H. B. Synopsis of the Foraminifera of the Upper and Middle Lias of Somersetshire.—In Charles Moore's paper—On the Middle and Upper Lias of the South-west of England. <*Proc. Somerset Arch., and Nat. Hist. Soc.*, vol. xiii, p. 104, pls. i-iii. 1867.

BRADY, H. B. On Ellipsoidina, a New Genus of Forminifera, by Guiseppe Seguenza; with further notes on its structure and affinities. <*Ann., Mag. Nat. Hist.*, ser. 4, vol. i, p. 333, pl. xiii. 1868.

BRADY, H. B. Notes on the Foraminifera of Mineral Veins and the Adjacent Strata. <*Report Brit. Assoc.* (Exeter Meeting), pp. 381–382. Also further notes in Charles Moore's paper. 1869.
See under *Moore*.

BRADY, H. B., in Brady, Robertson, and Brady's paper. The Ostracoda and Foraminifera of Tidal Rivers. <*Ann., and Mag. Nat. Hist.*, ser. 4, vol. vi, p. 273, pls. xi, xii. 1870.

BRADY, H. B. Catalogue of British Foraminifera in Edinburgh Museum of Science and Art. Edinburgh, 1870.

BRADY, H. B. On *Saccammina carteri*, a new Foraminifer from the Carboniferous Limestone of Northumberland. <*Ann., and Mag. Nat. Hist.*, ser. 4, vol. vii, pp. 177–184. 1871. Also *Nat. Hist. Trans. Northd. and Durham*, vol. vii, p. 177, pl. xli.

BRADY, H. B. Memoirs of the Geological Survey of Scotland. Explanation of Sheet 23—Lanarkshire. Central Districts, 8 vo. 1873.

BRADY, H. B. On *Archaediscus Karreri*, a new type of Carboniferous Foraminifera. <*Brit. Assoc. Advan. Sci.* (Bradford Meeting) 1873, p. 76.
Abstract.

BRADY, H. B. On *Archaediscus Karreri*, a new type of Carboniferous Foraminifera. <*Ann., and Mag. Nat. Hist.*, ser 4, vol. xii, pp. 286-290, pl. vi. 1873. (Abstract Report Brit. Assoc. (Bradford Meeting.)

BRADY, H. B. On a True Carboniferous Nummulite. <*Ann., and Mag. Nat. Hist.*, ser. 4, vol. xiii, p. 222, pl. xii. 1874.

BRADY, H. B. A monograph of Carboniferous and Permian Foraminifera, (The Genus Fusulina excepted.) *Palaeon. Soc.*, vol. xxx, pp. 1-166, plates i-xii. 1876.

BRADY, H. B. On some Foraminifera from Loo Choo Islands. *Proc. R. Irish Acad*, ser 2, vol. ii, p. 589. Also *Quart. Journ. Micr. Sci.*, vol. xvi. new series, p. 405. 1876.

BRADY, H. B. Notes on a Group of Russian Fusulinæ. *Ann., and Mag. Nat. Hist.*, ser. 4, vol. xvii, p. 414, pl. xviii. 1876.

BRADY, H. B. Zittel's Handbook of Palæontology. <*Nature*, vol. xiv, pp. 445-447. 1876.

BRADY, H. B. Supplementary Note on the Foraminifera of the Chalk (?) of the New Britain Group *Geol. Mag.*, dec. II, vol. iv, p. 534. 1876.

BRADY, H. B. In Prof. E. P. Wright's Notes on Foraminifera. (Seychelle Islands and Cagliari.) *Ann., and Mag. Nat. Hist.*, ser. 4, vol. xix, p. 103. 1877.

BRADY, H. B. Rhizopoda reticularia, in Sir. G. S. Nares' Narrative of a Voyage to the Polar Sea during 1875-6 in H. M. Ships "Alert" and "Discovery," vol. ii, app. xiii, p. 295. 8 vo. London, 1878.

BRADY, H. B. On the Reticularian and Radiolarian Rhizopoda. (Foraminifera and Polycystina) of the North Polar Expedition of 1875, 1876. *Ann., and Mag. Nat. Hist.*, ser. 5, vol. i, p. 425, pls. xx, xxi. 1878.

BRADY, H. B. Notes on some of the Reticularian Rhizopoda of the Challenger Expedition. *Quart. Journ. Micros. Sci.*, vol. xix—xxi, new series.
   I. On new or little known Arenaceous types, vol. xix, p. 20, pls. iii.—v.
   II. Additions to the knowledge of Porcellanous and Hyaline types, and notes on Pelagic Foraminifera, vol. xix, p. 261, pl. viii.
   III. Classification, Further Notes on New Species, and Note on Biloculina Mud, vol. xxi, p. 31. 1879-81.

BRADY, H. B. Notes on Rhizopoda obtained from Capt. Markham's Soundings on the Shores of Novaya Zemlya. In Markham's *A Polar Reconnaissance*, p. 346. 8 vo. London, 1881.

BRADY, H. B. Notes on some of the Reticularian Rhizopoda of the "Challenger" Expedition, by H. B. Brady, F. R. S. (*Microsc. Journal*, vol. xix, new ser.) *Bull., de la Soc. Belg., de Micros.*, vol. vi, pp. xvii-xxv. 1882.

BRADY, H. B. Note on the Keramosphæra, a new Type of Porcellanous Foraminifera. <*Ann., and Mag. Nat. Hist.*, ser. 5, vol. x, pp. 242-245. 1882.

BRADY, H. B. Pliocene Foraminifera, in Clement Reid's Geology of the Country around Cromer. <*Mem. Geol. Survey—England and Wales.* (Explanation of Sheet 68 E), p. 65. 1882.

BRADY, H. B. In the Exploration of the Faroe Channel, during the Summer of 1880, in H. M.'s hired ship "Knight Errant." By Staff Commander Tizard, R. N., and John Murray. *Proc. Roy. Soc.*, vol. xi, pp. 638-720. 1882.

See under Tizard and Murray.

BRADY, H. B. Note on Syringammina, a new type of Arenaceous Foraminifera. <*Proc. Roy. Soc.*, vol. xxxv, pp. 155-161, pl. ii, iii. 1883.

BRADY, H. B. Report on the Foraminifera dredged by H. M. S. Challenger, during the years 1873-1876. <*Report of the Scientific Results of the Voyage of H. M. S. Challenger during the years* 1873-1876. Zoology—vol. ix, text and plates, 2 parts. Folio. London, 1884.

BRADY, H. B., W. K. PARKER, and R. T. JONES. Monograph of the Genus Polymorphina. *Trans. Linn. Soc. Lond.*, vol. xxvii, p. 197, 4 pls. 1870.

BROWN, J. Notes on the Artesian Well at Colchester, and remarks on some of the Microscopic Fossils from the Colchester Chalk. *Ann., and Mag. Nat. Hist. Lond.*, ser. 2, vol. xii, p. 240, vii, ix. 1853.

BROWN, (Capt.) T. Conchologist's Text-Book, embracing the arrangements of Lamarck and Linnæus with a glossary of technical terms. Glasgow, 1833.

BROWN, (Capt.) T. Illustrations of the Conchology of Great Britain and Ireland. Edinburgh, 1827. Second edition, London, 1839.

BROCKLESBY, J. Views of the Microscopic World, pp. 54-58. 1851.

BRODIE, REV. P. B. Remarks on the Lias at Fretherne near Newnham, and Purton near Sharpness, with an account of some new Foraminifera discovered there. <*Ann., and Mag. Nat. Hist. London*, ser. 2, vol. xii, p. 272. 1853.

BROOKES, S. An introduction to the study of Conchology. Chap. xxii, pp. 91-96. (Nautilus. genus, xix.) 1815.

BRYCE, J. On the Order of Succession in the Drift-beds of the Island of Arran. <*Quart. Journ. Geol. Soc. Lond.*, vol. xxi, pp. 204-213. 1865.

BUCKLAND, (Rev.) DR. W. Notice of the Discovery of Fossil Foraminifera in the Mountain-Limestone of England in 1839 by Messrs. Tennant and Darker. <*Abstracts of the Proceedings of the Ashmolean Society*, vol. i, (Reprinted, *Edin. New Phil. Journ.*, vol. xxx, p. 441. 1841.

CARPENTER, W. B. On the Microscopic Structure of *Nummulina Orbitolites*, and Orbitoides. <*Quart. Journ. Geol. Soc. Lond.*, vol, vi, p. 21-38. 1849.

CARPENTER, W. B. Researches on the Foraminifera; Part I, General Introduction and Monograph of the Genus Orbitolites. <*Ann. Mag. Nat. Hist.*, vol. xvi, p. 207. *Amer. Journ. Sci.*, vol. xxi, 2d ser., pp. 429-432. 1856.

CARPENTER, W. B. The Microscope and its Revelations; with an appendix by Francis Gurney Smith, L. I..

    Chap. x. Foraminifera, Polycystina, and Sponges, pp. 436-456. Composition of Marine deposits pp. 631-634. Structure of *Nummulite* pp. 634-636. *Orbitoides* 637-639, Philadelphia, 1856.

CARPENTER, W. B. "Researches on the Foraminifera;" Part II. <*Proc. Roy. Soc.*, vol. viii, pp. 205-208. 1857. (Abstract.)

CARPENTER, W. B. Researches on the Foraminifera.—Part III. On the Genera *Peneroplis*, *Operculina*, and *Amphistegina*. <*Proc. Roy. Soc.*, vol. ix, pp. 334-337. 1859. (Abstract.)

CARPENTER, W. B. Researches on the Foraminifera <*Phil. Trans.* 1856-1860.

    1st Series.—Introduction; Genus Orbitolites, 1856, p. 181, pls. iv, ix.
    2d Series.—Genera Orbiculina, Alveolina, Cycloclypeus, and Heterostegina, 1856, p: 547, pls. xxviii.—xxxi.
    3d Series.—Genera Peneroplis, Operculina, and Amphistegina 1858, p. 1, pls. i.—vi.
    4th Series.—Genera Polystomella, Calcarina, Tinoporus, and Carpenteria. Concluding Summary, 1860, p. 535, pls. xvii.—xxii.

CARPENTER, W. B. General Results of the Study of Typical Forms of Foraminifera, in their Relation to the Systematic Arrangement of that Group, and to the Fundamental Principles of Natural History Classification. <*Nat. Hist. Rev.*, vol. i, pp. 185-201. 1861.

CARPENTER, W. B. On the Systematic Arrangement of the Rhizopoda. <*Nat. Hist. Rev.*, vol. i, pp. 456-472. 1861.

CARPENTER, W. B. Preliminary Report, of Dredging Operations in the Seas to the North of the British Islands, carried on in Her Majesty's steam vessel "Lightning," by Dr. Carpenter and Dr. Wyville Thomson. *Proc. Roy. Soc*, vol. xvii, pp. 168-200. 1868.

CARPENTER, W. B. On the Shell Structure of *Fusulina*. *Monthly Micr. Journ.*, 1869, p. 180. pls. xiv. 1869.

CARPENTER, W. B. The Geological Bearings of Recent Deep-sea Explorations. <*Nature*, vol. ii, pp 513-515. 1870.

CARPENTER, W. B. On the Rhizopodal Fauna of the Deep-sea. *Proc. Roy. Soc.*, vol. xviii, pp. 59-62. 1870.

CARPENTER, W. B. Descriptive Catalogue of Objects from Deep-sea Dredgings, exhibited at the Soiree of the Royal Microscopical Society, King's College, April 20, 1870. 12mo. London.

CARPENTER, W. B. Remarks on Professor Wyville Thomson's Preliminary Notes on the Nature of the Sea-bottom procured by the soundings of H. M. S. Challenger. <*Nature*, vol. xi, pp. 297-298. 1875. (Abstract.)

CARPENTER, W. B. Remarks on Professor Wyville Thomson's Preliminary Notes on the Nature of the Sea-bottom procured by the Soundings of H. M. S. "Challenger.". <*Proc. Roy. Soc.*, vol. xxiii, pp. 234-245. 1875.

CARPENTER, W. B. The Microscope and its Revelations. 5th ed. 12mo. London, 1875.

CARPENTER, W. B. On the conditions which determine the Presence or Absence of Animal Life on the Deep-sea Bottom. *Geol. Mag.*, new series, vol. ii, pp. 88-90. 1875.

CARPENTER, W. B. On the Origin of the Red Clay found by the "Challenger" at great Depths of the Ocean. <*Report Brit. Assoc.* (Bristol meeting). Trans. Sections, p. 64. 1875.

CARPENTER, W. B. Remarks on Mr. Carter's paper On the Polders. <*Ann. and Mag. Nat. Hist.*, ser. 4, vol. xvii, pp. 380-387. 1876.

CARPENTER, W. B. Art.—Foraminifera. <*Encyclopædia Britannica*, 9th ed., vol. ix, p. 371. 1871.

CARPENTER, W. B. On the Genus Astrorhiza of Sandahl, lately described as Haeckelina by Dr. Bessels. <*Quart. Journ. Micr. Soc.*, new series, vol. xvi, p. 221, pl. xix. 1876.

CARPENTER, W. B. The Microscope and its revelations.
Foraminifera and *Radiolaria* chapter xii, pp. 543-609. (Sixth edition.) 1881.

CARPENTER, W. B. Researches on the Foraminifera—Supplemental Memoir. On an Abyssal type of the genus Orbitolites; a Study in the Theory of Descent. *Phil. Trans.*, vol. clxxiv, pp. 551-573; pls. xxxvii, xxxviii. 1883.

CARPENTER, W. B. Report on the specimens of the Genus *Orbitolites*, collected by H. M. S. Challenger, during the years 1873-76. <*Zool. Chall. Exp.*, vol. vii, pp. 47. 8 plates. 1883.

CARPENTER, W. B., and H. B. BRADY. Description of Parkeria and Loftusia, two gigantic types of Arenaceous Foraminifera. <*Phil. Trans.*, 1869, p. 721; pl. lxxii-lxxx. 1869.

CARPENTER, W. B., and JEFFREYS, Dr. J. GWYN. Report on Deep-sea Researches, carried on during the months of July, August and September, 1870, in H. M. Surveying Ship "Porcupine." <*Proc. Roy. Soc.*, vol. xix, p. 146, 1870.

CARPENTER, W. B., W. K. PARKER, and T. R. JONES. Introduction to the Study. 319 pages. Fol., 22 plates. *Roy. Society* 1862.

CARRUTHERS, W. On Traquairia, a Radiolarian Rhizopod from the Coal-measures. <*Report Brit. Assoc.*, (Brighton meeting), Trans. Sec., p. 126. 1872.

CARTER, H. J. On the Structure of the larger Foraminifera. <*Ann. and Mag. Nat.*, vol. viii, p. 246. 1861.

CARTER, H. J. On two New Species of the Foraminiferous Genus Squamulina; and on a New Species of Difflugia. <*Ann. and Mag. Nat. Hist.*, ser. 4, vol. v, pp. 309-326; pls. iv, v. 1870.

CARTER, H. J. Notes on the Branched Variety of Squamulina scopula. <*Ann. and Mag. Nat. Hist.*, ser. 4, vol. vi, p. 346. 1870.

CARTER, H. J. On Haliphysema ramulosa (Bowerbank) and the Sponge-spicules of Polytrema. <*Ann. and Mag. Nat. Hist.*, ser. 4, vol. v, p. 389. 1870.

CARTER, H. J. On Melobesia unicellularis, better known as the *Coccolith*. <*Ann. and Mag. Nat. Hist*, ser. 4, vol. vii, pp. 184-189. 1871.

CARTER, H. J. Absence of microscopic calcareous Organic Remains in Marine Strata charged with siliceous ones. <*Nature*, vol. xi, p. 186. 1875.

CARTER, H. J. On the Locality of Carpenteria balaniformis, with Description of a new Species, and other Foraminifera found in and about Tubipora musica. <*Ann., and Mag. Nat. Hist.*, ser. 4, vol. xix, p. 209, pl. xiii. 1877.

CARTER, H. J. On the branched form of the Apertural Prolongation from the summit of Carpenteria monticularis. <*Ann., and Mag. Nat. Hist.*, ser. 4, vol. xx, p. 68, wood cut. 1877.

CARTER, H. J. On a Melobesian form of Foraminifera (Gypsina melobesioides, mihi); and further observations on Carpenteria monticularis. <*Ann., and Mag. Nat. Hist.*, ser. 4, vol. xx, p. 172. 1877.

CARTER, H. J. Description of a new Species of Foraminifera (Rotalia spiculotesta). <*Ann., and Mag. Nat. Hist.*, ser. 4, vol. xx, p. 470, pl. xvi. 1877.

CARTER, H. J. On Stromatopora. *Ann, and Mag. Nat. Hist.*, ser. 5, vol. ii, p. 85. 1878.

CARTER, H. J. Position of the Sponge spicule in the Spongidæ; and Postscript on the identity of Squamulina scopula with the Sponges. <*Ann, and Mag. Nat. Hist.*, ser. 5, vol. i, pp. 170-174. 1878.

CARTER, H. J. On a New Genus of Foraminifera (*Aphrosina informis*) and Spiculation of an unknown Sponge. <*Jour. R. Micr. Soc. Lond.*, vol. ii, pp. 500-502, 1 pl. 1879.

CARTER, H. J. Notes on Foraminifera. <*Ann., and Mag. Nat Hist.*, ser. 5, vol. iii, p. 407. 1879.

CARTER, H. J. Report on specimens dredged up from the Gulf of Manaar and presented to the Liverpool Free Museum by Capt. W. H. Cawne Warren. <*Ann.*, *and Mag. Nat. Hist.*, ser. 5, vol. v, p. 437, pls. xviii, xix. 1880.

CARTER, H. J. Supplementary Report on Foraminifera and Sponges, Dredged up from the Gulf of Manaar, together with others from the Sea in the vicinity of the Basse Rocks and from Bass's Straits, presented to the Liverpool Free Museum by Captain W. H. Cawne Warren. <*Proc. Lit. and Philo. Soc. Liv.*, vol. xxxv, pp. 271-275. 1881. Same reprinted in the *Ann., and Mag. Nat. Hist.*, ser. 5, vol. vii, pp. 361-385, pl. xviii, 1881.

CARTER, H. J. Note on the assumed Relationship of Parkeria to Stromatopora, and on a Microscopic Section of Stromatopora mamillata, Fr. Schmidt. <*Ann., and Mag. Nat. Hist.*, ser. 5, vol. xiii, pp. 353-356. 1884.

CARTER, H. J. Remarks on Prof. Haeckel's Observations on Wyvillethomsonia Wallichii and Squamulina scopula. <*Ann., and Mag. Nat. Hist.*, ser. 4, vol. xx, pp. 337-339. 1877.

CARTER, H. J. Points of Distinction between the Spongiadae and the Foraminifera. <*Ann., and Mag. Nat. Hist.*, ser. 4, vol. xi, pp. 351-356. 1873.

CARTER, H. J. On the Striæ of Foraminiferous Tests, with a reply to Criticism. <*Ann., and Mag. Nat. Hist.*, ser. 4, vol. xiv, p 138. 1874.

CARTER, H. J. On the Polythemata (Foraminifera) especially with reference to their Mythical Hybrid Nature. <*Ann., and Mag. Nat. Hist.*, ser. 4, vol. xvii, p. 185, pl. xiii. 1876.

CARTER, H. J. Parkeria inferred to have been a species of Hydractinia. <*Ann., and Mag. Nat. Hist.*, ser. 4, vol. xviii, p. 187. 1876.

CARTER, H. J. On the close Relationship of Hydractinia, Parkeria, and Stromatopora, with Descriptions of new Species of the former, both Recent and Fossil. <*Ann., and Mag. Nat. Hist.*, ser. 4, vol. xix, p. 44, pl. viii. 1877.

CARTER, H. J. Description of Bdelloidina aggregata, a new genus and species of Arenaceous Foraminifera, in which their so-called "Imperforation" is questioned. *Ann., and Mag. Nat. Hist.*, ser. 4, vol. xix, p. 201, pl. xiii. 1877.

CHIMMO, WM. Bed of the Atlantic Ocean, in latitude 47° N., longitude 23° W., are taken upwards of 100 (Microscopic Drawings of) Minute Organisms, 40 n. p. 1870.
   Not seen.

CLARK, WM. Observations on the recent Foraminifera. <*Ann. and Mag. Nat. Hist.*, ser. 2, vol. v, p. 380. 1849.

CLARK, WM. On the recent Foraminifera. <*Ann. and Mag. Nat. Hist.*, ser. 2, vol. v, p. 161. 1850.

COCKS, W. P. Contributions to the Fauna of Falmouth. Foraminifera. <*Seventeenth Ann. Report Roy. Cornwall, Pol. Soc.*, p. 87. 1849

COSTA, E. da., Mendes, Historia Naturalis Testaceorum Brittaniæ. London, 1778.

CROSSKEY, H. W., and D. ROBERTSON. The Post-tertiary Fossiliferous Beds of Scotland, parts I-XX. <*Trans. Geol. Soc Glasgow*, vol. ii, p. 267;—vol. iii, pp. 113, 321;—vol. iv, pp. 32, 128 and 241;—vol. v, p. 29. 1867-1876.

CROUCH, E. A. An Illustrated Introduction to Lamarck's Conchology; contained in his Histoire Naturelle des Animaux sans Vertebres: being a literal translation of the descriptions of the Recent and Fossil Genera, p. 47, pl. 22. London, 1827.

DAWSON, J. W. Acadian Geology, 2d ed., 8vo. London, 1868.

DEANE, H. On the occurrence of Fossil Xanthidia and Polythalmia in Chalk. <*Trans. Micr. Soc. Lond*, vol. ii, pp. 77-79. 1845.

DIXON, F. The Geology and Fossils of the Tertiary and Cretaceous Formations of Sussex. London, 1850.

DUNCAN, P. M. Note on the Scindian Fossil Corals. <*Quart. Geol. Soc. Lond.*, vol. xx, pp. 66-72. 1864.

DUNCAN, P. M. On the Syringosphæridæ, an Order of Extinct Rhizopoda. <*Ann. and Mag. Nat. Hist.*, ser. 5, vol. ii, pp. 297-299. 1878.

DUNCAN, P. M. On the genus Stoliczkaria, Dunc., and its distinctness from Parkeria, Carp. <*Quart Jour. Geol. Soc. Lond.*, vol. xxxviii, pp. 69-74, pl. ii. 1882.

ELCOCK, C. Foraminifera at Southport. <*Journ. Postal Micr. Soc.*, vol. ii, p. 120. 1882.

ELCOCK, C. Preparing Fossil Foraminifera, Spicula, etc. <*Journ. Micr. Soc., Lond.*, n. s., vol. ii, pp. 886, 887 1882.

ELCOCK, C. List of Foraminifera from Silt. <*Journ. Postal Micr. Soc.*, vol. ii, pp. 119, 120. 1883.

ELCOCK, C. Notes on the Occurrence of some rare Foraminifera in Irish Sea. <*Ann. and Mag. Nat. Hist.*, ser. 5, vol. xiv, pp. 366, 367. 1884.

ELEY, (Rev.) H. Geology in the Garden; or the Fossils in the Flint Pebbles. 8 vo. London, 1859.

ETHERIDGE, R. (Jun.) On the Occurrence of Foraminifera (Saccammina Carteri, Brady), in the Carboniferous Limestone Series of the East of Scotland. *Trans. Edin. Geol Soc.*, vol. ii, pp. 225, 236. 1873.

ETHERIDGE, R. (Jun.) Note on the Fossils from the Glacial Deposits of the North-west Coast of the Island of Lewis, Outer Hebrides *Geol. Mag.*, n. s., dec. II, vol. iii, p. 552. 1876.

FLEMING, J. Philosophy of Zoology; or, a General view of the Structure, Functions, and Classification of Animals. 8 vo. Edinburgh. 1822.

FLEMING, J. Observations on some Species of the Genus Vermiculum of Montague. <*Mem. Wern. Soc.*, vol 4, part ii, pp. 564-567; pl. xv. 1823.

FLEMING, J. A History of British animals, exhibiting the descriptive characters and systematic arrangement of the genera and species of Quadrupeds, Birds, Reptiles, Fishes, Mollusca, and Radiata of the United Kingdom. 8 vo. Edinburgh, 1828.

FOLIN, M. de. On a new State of Reticularian Rhizopods. <*Ann. and Mag. Nat. Hist.*, ser. 5, vol. xvi, pp. 232, 233. 1882.

GEDDES, P. On the Nature and Functions of the "Yellow Cells" of Radiolarians and Coelenterates· <*Proc. Roy. Soc. Edin*, vol. xi, pp. 377-396. 1882.

GOSSE, P. H. Rhizopoda (Foraminifera), Marine Zoology. Part I, pp. 8-14. 1855.

GOSSE, P. H. On the Presence of Motile Organs, and the Power of Locomotion, in Foraminifera. <*Ann. Mag. Nat. Hist.*, ser 2, vol. xx, pp. 365-367. 1857.

GRAY, J. E. On Carpenteria and Dujardinia, two genera of a new form of Protozoa, with attached multilocular Shells filled with Sponge apparently intermediate between Rhizopoda and Porifera. <*Proc. Zool. Soc. Lond.*, vol. xxvi, p. 226, wood cuts. 1858.

GREEN, J. Foraminiferous Silt Banks of the Isle of Ely. <*Journ. Roy. Mic. Soc. Lond.*, n. s., vol. i, p 473. 1881.

GUMBEL, C. W. On Deep-Sea Mud. <*Nature*, vol. iii, pp. 16, 17. 1870.

GUPPY, R. L. J. The Origin of Coral Reefs. Shortlands Islands, Solomon Group." *Nature*, vol. xix, pp. 214, 215. 1884.

HARDMAN, E. T. The Deep-Sea Manganiferous Muds. <*Nature*, vol. xv, pp. 57, 58. 1877.

HOEVEN, J. VAN DER. Systematic arrangement of Infusories. <Hand-Book of Zoology, vol. i, pp. 45-59. 1856.

HOOKE, R. Micrographia; or some Physiological Descriptions of Minute bodies made by Magnifying Glasses, with Observations and Inquiries thereupon. Thirty-eight plates, folio. London, 1665.

HOOKE, R. Micrographia, or some Physiological Descriptions of Minute Bodies, made by Magnifying Glasses, with Observations and Inquiries thereupon. Fol , 38 plates. London, 1667.

HOUSE, R., and J. W. KIRKLY. Synopsis of the Geology of Durham and part of Northumberland. 8 vo. Newcastle-on-Tyne. 1863.

HULL, E. Nature of the Oceanic Bed at great Depths. <*Jour. Roy. Geol. Soc. Ireland*, vol. iv, pp. 55-59. 1879.

HUXLEY, T. H. Upon *Thalassicolla*, a new Zoophyte. <*Ann. Mag. Nat. Hist*., ser. 2, vol. viii, pp. 433-442. 1851.

HUXLEY, T. H. Report on the Examination of Specimens of Bottom. In Report on Deep-Sea Soundings in the North-Atlantic Ocean between Ireland and Newfoundland, made in H. M. S. "Cyclops," Lieut. Comm. Joseph Dayman, in June and July 1857, p. 62, pl. iv. 1858.

HYNDMAN, G. L. Report of the Proceedings of the Belfast Dredging Committee. *Brit. Assoc. Advan. Sci.*, 1857, pp. 220-237. 1858.
List of Foraminifera p. 237.

JAMESON, R. Notes on the geology of the countries discovered during Captain Parry's second expedition, A. D. 1821-22-23. In Journal of a Third Voyage for the Discovery of a North West Passage from the Atlantic to the Pacific; performed in the years 1824-25, in His Majesty's ship Hecla and Fury, under the orders of Captain William Edward Parry." Appendix, pp. 132-151. London, 1826.
Contains various notes on the fossils observed during the expedition. Mr. Stokes communicates a note on a fossil from limestone of the island of Igloolik, which is clearly a species of *Receptaculites*.

JAMIESON, T. F. On the Pleistocene Deposits of Aberdeenshire. <*Quart. Journ. Geol. Soc. Lond.*, vol. xiv, pp. 509-532. 1858.

JAMIESON, —. Balanus, and some Foraminifera found in the Post-Tertiary shell-beds in Nairnshire. <*Trans. Edin. Geol. Soc.*, vol. iv, pp. 141, 142. 1882.

JEFFREYS, J. G. Notes on British Foraminifera. *Proc. Roy. Soc.*, vol. vii, No. 14. 1855.

JEFFREYS, J. G. Notes on British Foraminifera. <*Ann. and Mag. Nat. Hist.*, vol. xvi, p. 207. *Amer. Journ. Sci.*, vol. xxi, 2d ser., pp. 432-434. 1856.

JEFFREYS, J. G. Deep-Sea Exploration. <*Nature*, vol. xxiii, pp. 300-302, 324-326 1881.

JONES, T. R., in Prof. W. King's —— Monograph of the Permian Fossils of England. <*Palaeontographical Society's Monographs.* 1850.

J., T. R. On Recent Notices of the Nummulite Formation. By C. Giebel. *Quart. Journ. Geol. Soc. Lond.*, vol. vii, pp. 116-118. 1851.

JONES, T. R., in Prestwich's paper—On the Structure of the Strata between the London Clay and the Chalk, in the London and Hampshire Tertiary Systems. Part iii. The Thanet sands. <*Quart. Journ. Geol. Soc. Lond.*, vol. viii, p. 235; pls. xv-xvi. 1852.

JONES, T. R., in the Rev. P. B. Brodie's—Remarks on the Lias at Fretherne, near Newnham, and Purton near Sharpness, with an account of some new Foraminifera discovered there. *Ann. and Mag. Nat. Hist.*, ser. 2, vol. xii, p. 272. 1853.

JONES, T. R. Chalk Foraminifera, in S. J. Mackie's—Thoughts on Dover Cliffs. <*Geologist*, vol. vi; p. 293, and p. 432. 1863.

Jones, T. R. In Prof. Prestwick's Anniversary Address. <*Quart. Journ. Geol. Soc. Lond* , vol. xxvii, p. 51. 1871.

Jones, T. R. On the Range of the Foraminifera in Time. <*Proc. Geologists' Assoc.*, vol. ii, p. 187. 1872.

Jones, T. R. On some Foraminifera in the chalk of the North of Ireland. <*Journ. Roy. Geol. Soc.*, n. s., vol. iii, pp. 88–91. 1873.

Jones, T. R. On Quartz and other Forms of Silica. <*Nature*, vol. xiii, pp. 159–160. 1875.
(Foraminifera.)

Jones, T. R. In Griffith and Henfrey's Micrographic Dictionary, 3d ed., 8vo. London, 1875.

Jones, T. R. Oolitic Foraminifera of England. <In Phillips's *Geology of the Yorkshire Coast*, 3d ed., p. 278. 1875.

Jones, T. R. Remarks on the Foraminifera, with special reference to their Variability of Form, illustrated by the Cristellarians. <*Monthly Micr. Journ.*, vol. xv, p. 61, pls. cxxviii, cxxix.
Note on Prof. Rupert Jones's Memoir on the variability of Foraminifera. Ibid. p. 200. 1876.

Jones, T. R. The Late Prof. Ch. G. Ehrenberg's Researches on the Recent Foraminifera. <*Monthly Micr. Journ.*, vol. xvii, p. 300;—vol. xviii, p. 49. 1877–78.

Jones, T. R. In Dixon's Geology of Sussex, new ed., pt. ii, p. 168, etc. 1878.

Jones, T. R. Note on the Foraminifera and other Organisms in the Chalk of the Hebrides. <*Quart. Journ. Geol. Soc. Lond* , vol. xxxiv, pp 739, 740. 1878.

Jones, T. R. On some Foraminifera in the Chalk of the North of Ireland. <*Journ. R. Geol. Soc. Ireland*, vol. iii, p. 88. 1879.

Jones, T. R Catalogue of the Fossil Foraminifera in the British Museum. 8vo. London, 1882.

Jones, T. R. The importance of Minute things of Life in Past and Present times. <*Trans. Hert. Nat. Hist. Soc. and Field Club*, vol. ii, pt. 4, pp. 164–172, 1883.

Jones, T. R. Notes on the Foraminifera and Ostracoda from the Deep Boring at Richmond. <*Quart. Journ. Geol. Soc. Lond.*, vol. pp. 765–777, plate xxxiv. 1884.

Jones, T. R. The Origin and Composition of Chalk and Flint, with special reference to their Foraminifera and other Minute Organisms. <*Trans. Hert. Nat. Hist. Soc. F. Club.*, vol iii, pp. 143–156. 1885.

Jones, T. R., W. K. Parker and H. B. Brady. A Monograph of the Foraminifera of the Crag. <*Palaeon. Soc.*, vol. xix, pp. 1–72, 3 tables, 4 plates. 1865.

JONES T. R. and W. K. PARKER. On the Rhizopodial Fauna of the Mediterranean, compared with that of the Italian and some other Tertiary Deposits. <Quart. Journ. Geol. Soc. Lond., vol. xvi, pp. 292-307. 1860.

JONES T. R. and W. K. PARKER. On some Fossil Foraminifera from Chellaston near Derby. <Quart. Journ. Geol. Soc. Lond., vol. xvi, pp. 452-458, 2 plates. 1860.

JONES T. R. and W. K. PARKER. On the Foraminifera of the Crag. <Ann., and Mag. Nat. Hist., ser. 3, vol. xiii, p. 64. 1864. See also Mem. Geol. Survey of Gt. Britain, Geology. Middlesex, etc., p. 59.

JONES T. R. and W. K. PARKER. On the chalk of Gravesend and Mendon, figured by Prof. Dr. Chr. G. Ehrenberg (in 1854). <Geol. Mag., new series, vol. viii, p. 506. 1871.

JONES T. R. and W. K. PARKER. On the Foraminifera of the Family Rotalinae (Carpenter), found in the Cretaceous Formation, with Notes on their Tertiary and Recent Representatives. <Quart. Journ. Geol. Soc. Lond., vol, xxvii, pp. 103-131. 1872.

JONES T. R. and W. K. PARKER. Notes on Eley's Foraminifera from the English Chalk. <Geol. Mag., vol. ix, p. 123. 1872.

JONES T. R. and W. K. PARKER. Lists of some English Jurassic Foraminifera. <Geol. Mag , dec. II, vol. ii, pp. 308-311. 1875.

JONES T. R. and W. K. PARKER. On some Recent and Fossil Foraminifera dredged up in the English Channel. <Ann., and Mag. Nat. Hist., ser. 4, xvii, p. 283, wood cuts. 1876.

JUDD J. W. and C. HOMERSHAM. Supplementary Notes on the Deep Boring at Richmond, Surrey. Quart. Journ. Geol. Soc., vol. xli, pp. 523-528. 1885.

KANMACHER, F. Adams's Essays on the Microscope; the second edition, with considerable additions and improvements. 4to, with folio plates. London, 1798.

KINAHAN, G. H. On the Cretaceous Period. Nature, vol. iii, p 286. 1871.

KEEPING, W. On some Remains of Plants, Foraminifera and Annelida in the Silurian Rocks of Central Wales. Geol. Mag., dec. II, vol. ix, pp. 485-491, pl. xi. 1882.

KENT, W. S. The Foraminiferal Nature of Haliphysema Tumanowiczie, Bow. (Squamulina scopula, Carter) demonstrated. Ann and Mag. Nat. Hist., ser. 5, vol. ii, p. 68, pls iv, v. 1878.

KENT, W. S Observations of Professor Ernst Haeckel's Group of the Physemaria, and on the Affinity of the Sponges Ann. and Mag. Nat. Hist , ser. 5, vol. i, p. 1-17. 1878.

KING, W. A. Catalogue of the Organic Remains of the Permian Rocks of Northumberland and Durham. 8 vo. Newcastle-on-Tyne, 1848.

KING, WM. A Monograph of the Permian Fossils of England. London, 1850.

KING, W. On the Occurrence of Permian Magnesian Limestone at Tullyconnel, near Artrea, in the County of Tyrone. <*Journ Geol. Soc.*, Dublin, vol. vii, part 2. 1850.

KING, W. Oceanic Sediments and their Relation to Geological Formations. <*Ann. and Mag. Nat. Hist.*, ser. 4, vol. xv, pp. 198–204. 1875.

KING, W., and T. H. ROWNEY. An old chapter of the Geological Record. 8 vo. London, 1881.

KIRKLY, J. W. Brachiopoda, Polyzoa, and Foraminifera from the Permian Rocks of South Yorkshire. <*Quart. Journ. Geol. Soc*, *Lond.*, vol. xvii, pp. 306–309. 1861.

LAMPLUGH, G. W. On the Bridlington and Dimlington Glacial Shell-beds. <*Geol. Mag.*, dec. II, vol. viii, p. 535. 1881.

LANKESTER, E. R. The Structure of Haliphysema Tumanowiczii. <*Quart. Journ. Micr. Soc.*, vol. xix, new ser., p. 475, pl. xxii. 1879.

LATHAM, A. G. On Foraminifera from Dogs Bay, Roundstone, and from Berwick Bay. <*Proc. Lit. Philos. Soc*, *Manchester*, vol. vi, pp 85, 191, 1867.

LEBOUR, G. A. On the "Great" and "Four-fathom" Limestone and their associated beds in South Northumberland. <*Trans. N. of Eng. Inst. Min. Engineers*, vol. xxiv. 1875.

LEBOUR, G. A. Range of Saccummina Carteri (Brady). <*Geol. Mag.*, n. s., dec. II, vol. iii, p. 135. 1876.

LEGG, M. S. Observations on the Examination of Sponge Sand, with remarks on collecting, mounting, and viewing Foraminifera as microscopic objects. <*Quart. Journ. Micro. Sci.*, vol. i, 1853. Also, *Trans. Micro. Soc.*, *Lond.*, ser. 2, vol. ii, pl. xix.

LINTON, J. On a Sample of Sand from Dogs Bay, Connemara, skimmed from the Surface of the Sea. <*Proc. Lit. Philos. Soc.*, *Manchester*, vol. vi, pp. 184–186. 1867.

LISTER, M. Historiæ animalium Angliæ tres tractatus; Unus de Araneis, Alter de Cochleis tum terrestribus tum fluviatilibus, Tertius de Cochleis marinis, etc., cum Tab. œn. ix. Londini. 1678.

LIVERSIDGE, A. On the Occurrence of Chalk in the New Britain Group. <*Geol. Mag.*, n. s., dec II, vol. iv, p. 539. 1877.

MAC COY, F. Contributions to British Palæontology, 1854.

MACDONALD, J. D. Further Observations on deep soundings obtained by H. M. S. "Herald," Capt. Denham, employed on the Surveying Service in South-western Pacific. <*Ann. and Mag. Nat. Hist.*, ser. 2, vol. xxi. 1857.

MACGILLIVRAY, W. A History of the Molluscous Animals of the counties of Aberdeen, Kincardine, and Banff, &c. 12mo. London, 1843.

MACKIE, S. J. Microscopic Geology. <*Recreative Science.*, vol. i, pp. 145-150. 1860.

MANTELL, G. A. Thoughts on Animalcules. 12mo. 1846.

MANTELL, G. A. The soft bodies of Polythalmia found in fossil state. <*Trans. of the Roy. Soc. of Lond.* and in *Amer. Jour. Sci.*, vol. ii. 1846.

MANTELL, G. A. On the Fossil Remains of the soft parts of Foraminifera, in the Chalk and Flint of the Southeast of England. <*Amer. Journ. Sci.*, vol. v. 2d ser., pp. 70-74. 3 wood cuts. 1848.

MANTELL, G. A. "On the Fossil Remains of the soft parts of Foraminifera discovered in the Chalk Flint of Southeast of England." <*Proc. Roy. Soc.*, vol., v, pp. 627, 628. 1851.

MANTELL, G. A. Pictorial Atlas of Fossil Remains. Plates 61, 62. 1850.

MANTON, W. G., and RACKETT, REV. T. A Descriptive Catalogue of the British Testacea. <*Trans. Linnean Soc.*, vol. viii. 1807.

MCANDREW, R. List of the British Marine Invertebrate Fauna. Pp. 234, 235, (Foraminifera). <*Brit. Assoc. Advan. Sci.*, 1860-1861.
_{This list of British Foraminifera is taken from Prof. Williamson's "Recent Foraminifera of Great Britian," published by the Ray Society.}

M'COY, F. On some new Genera and Species of Palæozoic Corals and Foraminifera. <*Ann. and Mag. Nat. Hist.*, ser. 2, vol. iii, p. 131. 1849.

MEASURES, J. W. Foraminifera from Silt. <*Journ. Micr. Soc.*, vol. xiv, pp. 118, 119. 1883.

MIVART, (SR.) G. Notes touching Recent Researches on the Radiolaria. *Journ. Linn. Soc.*, vol. xiv, pp. 136-186, 16 wood cuts. 1878.

MOORE, C. On the Abnormal Conditions of Secondary Deposits when connected with the Somersetshire and South Wales Coal-Basin, and on the age of the Sutton and Southerndown Series. <*Quart. Journ. Geol. Soc. Lond.*, vol. xxiii, pp. 449-568, 2 plates. 1867.

MOORE, C. Report on Mineral Veins in Carboniferous Limestone and their Organic Contents. *Report Brit. Assoc.* (Exeter Meeting) pp. 360-388. 1869.

MOORE, C. On the Palæontology and Physical Conditions of the Meux-Well. *Quart. Jour. Geol. Soc.*, vol, xxxiv, p. 914. 1878.

MORRIS. J. Catalogue of British Fossils. London, 1843. (2d Edit. 1852.)

MORRIS, J., and J. QUEKETT. Catalogue of the Hunterian Museum of the Royal College of Surgeons of England, p. 87. 4to. London, 1860.

MOSELEY, H. N. Pelagic Life. *Nature*, vol. xvi, pp. 559-564. 1882.

MONTAGU, G. Testacea Britannica, or Natural History of British Shells, Marine, Land, and Fresh-water, etc. 3 vols. 4o. London, 1803.

MONTAGU, G. A supplement to the Testacea Britanica. London, 1808.

MOSELEY, H. N. Notes by a Naturalist on the "Challenger," being an account of various observations made during the voyage of H. M. S. "Challenger" round the world, in the years 1872–1876. London, 1879.

MUNIER-CHALMAS, and C. SCHLUMBÉRGER. New Observations on the Dimorphism of the Foraminifera. <*Ann., and Mag. Nat. Hist.*, ser. 5, vol. xl, pp. 336–340. 1883.

MURRAY, J. Preliminary Reports to Professor Wyville Thompson F. R. S., Director of the Civilian Scientific Staff, on Work done on board the "Challenger." <*Proc. Roy. Soc.*, vol. xxiv, pp. 471–544, 4 plates. 1876.

MURRAY, J. Deep-Sea Muds. <*Nature*, vol. xv, pp. 319, 340. 1877.

MURRAY, J. On the Structure and Origin of Coral Reefs and Islands (Abstract). <*Proc. Roy. Soc. Edinb.*, vol. x. pp. 505–518. 1880.

NEEDHAM, T. V. An Account of some new Microscopical Discoveries, plate 6. London, 1745.

NEVILL, T. H. Foraminifera from a deposit at Montreal. <*Proc. Lit., and Phil. Soc.*, Manchester, vol. iii, p. 100. *Quart. Journ. Micr. Sci.*, vol. iii, n. s., p. 211. 1863.

NICHOLSON, H. A, and R. ETHERIDGE, Jun. A Monograph of the Silurian Fossils of the Girvan District in Ayrshire, with especial reference to those contained in the "Gray Collection," fasc. i. 1878.

NORMAN, A. M. In Jeffreys and Norman's Submarine Cable Fauna. <*Ann., and Mag. Nat. Hist.*, ser. 4, vol. xv, p. 169, pl. xii. 1875.

NORMAN, A. M. In Dr. Jeffrey's Preliminary Report of the Biological Results of a Cruise in H. M. S. "Valorous" to Davis Strait in 1875. Proc. Roy. Soc., vol. xxv, p. 202. Also Dr. W. B. Carpenter. Ibid, p. 223. 1876.

NORMAN, A. M. Notes on the French Exploration of Le "Travailleur" in the Bay of Biscay (Abstract). <*Report Brit. Assoc.* (Swansea Meeting) p. 387. 1880.

NORMAN, A. M. On the Genus Haliphysema, with a description of several forms apparently allied to it. <*Ann., and Mag. Nat. Hist.*, ser. 5, vol. i, p. 265, pl. xvi. 1878.

NORMAN, A. M. On the Architectural Achievements of little Masons, Annelidan and Rhizopodan, in the Abyss of the Atlantic. <*Ann., and Mag. Nat. Hist.*, ser. 5, vol. i, p. 284. 1878.

NORMAN, A. M. Presidential Address. Part. II. The Abysses of the Ocean. <*Nat. Hist. Trans. Northd. and Durham*; vol viii, p. 25. 1883.

NORTHAMPTON (Marquis of). On Spirolinites in Chalk and Chalk-flints. <*Lond. and Edin. Phil. Mag.*, also *Proc. Geol. Soc. Lond*, vol. ii, p. 685. 1838.

OWEN, S. R. J. On the Surface fauna of Mid-Ocean. <*Journ. Linn. Soc. Lond.*, (Zoology) vol. ix., p. 147, pl. v. 1867.

PARFITT, E. On the Protozon of Devonshire. <*Trans. Devon. Assoc. Sci. Lit. and Art.*, vol. iii. 1869.

PARFITT, E. On a Species of Arenaceous Foraminifer? from the Carboniferous Limestone of Devonshire. <*Ann., and Mag. Nat. Hist.*, ser. 4, vol. vii, pp. 158-161. 1871.

PARFITT, E. On a new Species of Cellepora. <*Ann., and Mag. Nat. Hist.*, ser. 4, vol. xii, pp. 68, 69, pl. iii. B., 1873.

PARFITT, E. On the Structure of Haliphysema Tumanowiczii. <*Ann., and Mag. Nat. Hist.*, ser. 6, vol. ii, p. 88. 1878.

PARKER, W. K. and T. R. JONES, in Ansted's paper on Malaga,—Foraminifera of the Blue Clay of Tejares, Malaga. <*Quart. Journ. Geol. Soc., Lond.*, xv, p. 600. 1859.

PARKER, W. K., and T. R. JONES. On some Foraminifera from the North Atlantic and Arctic Oceans, including Davis Strait and Baffin Bay. <*Proc. Roy. Soc.*, vol. xix, pp. 239, 240. 1864.

PARKER, W. K., and T. R. JONES. On some Foraminifera from the North Atlantic and Arctic Oceans, including Davis Straits and Baffin's Bay. <*Phil. Trans. Roy. Soc. Lond.*, vol. clv, pp. 325-441; 7 plates. 1865.

PARKER, W. K., and T. R. JONES. On the Nomenclature of the Foraminifera, Part I, Linnæus and Gmelin. <*Ann. and Mag. Nat. Hist*, ser. 3, vol. iii, p. 474. 1859.

PARKER, W. K., and T. R. JONES. On the Nomenclature of the Foraminifera; Part II, Walker and Montagu. <*Ann. and Mag. Nat. Hist.*, ser. 3, vol. iv, p. 333. 1859.

PARKER, W. K., and T. R. JONES. On the Nomenclature of the Foraminifera. Part III, Fichtel and Moll. <*Ann. and Mag. Nat. Hist.*, ser. 3, pp. 98, 174. 1860.

PARKER, W. K., and T. R. JONES. On the Nomenclature of the Foraminifera Part IV, Lamarck. <*Ann. and Mag. Nat. Hist*, ser. 3, vol. v, pp. 285, 466; vol. vi, p. 29. 1860.

PARKER, W. K., and T. R. JONES. On the Nomenclature of the Foraminifera. Part V, De Montfort. *Ann. and Mag. Nat. Hist.*, ser. 3, vol. vi, p. 337. 1860.

PARKER, W. K., and T. R. JONES. On the Nomenclature of the Foraminifera. Part VI, Alveolina. *Ann. and Mag. Nat. Hist.*, ser. 3, vol. viii, p. 161. 1863.

PARKER, W. K., and T. R. JONES. On the Nomenclature of the Foraminifera. Part VII, Operculina and Nummulina. <*Ann. and Mag. Nat. Hist.*, ser. 3, vol. viii, p. 229. 1861.

PARKER, W. K., and T. R. JONES. On the Nomenclature of the Foraminifera. Part VIII.—Textularia. <*Ann. and Mag. Nat. Hist.*, ser. 3, vol. xi, pp. 91-98. 1863.

PARKER, W. K., and T. R. JONES. On the Nomenclature of the Foraminifera. Part IX. *The species enumerated by De Blainville and Defrance.* <*Ann. amd Mag. Nat. Hist.*, ser. 3, vol. xii, pp. 200-219. 1863.

PARKER, W. K , and T. R. JONES. On the Nomenclature of the Foraminifera. *The species enumerated by D'Orbigny in the "Annales des Sciences Naturelles," vol. vii*, 1826. <*Ann. and Mag. Nat. Hist.*, ser. 3, vol. xii, pp. 429-440. 1863.

PARKER, W. K., T. R. JONES, and H. B. BRADY. On the Nomenclature of the Foraminifera. Part XI.—*The species enumerated by Batsch in* 1791. *Ann. and Mag. Nat. Hist.*, ser. 3, vol. xv, pp. 225-232. 1865.

PARKER, W. K., T. R. JONES, and H. B. BRADY. On the Nomenclature of the Foraminifera. Part XII.—*The species enumerated by D'Orbigny* in the Annales des Sciences Naturelles, vol. vii, 1826. (3) The species illustrated by Models. <*Ann. and Mag. Nat. Hist.*, ser. 3, vol. xvi, p. 15, pls. i-iii. 1865.

JONES, T. R., W. K. PARKER, and J. W. KIRKBY. On the Foraminifera. Part XIII.—*The Permian Trochammina pusilla and its Allies.* <*Ann. and Mag. Nat. Hist.*, ser. 4, vol. iv, pp. 386-392. 1869.

PARKER, W. K., and T. R. JONES. On the Nomenclature of the Foraminifera. Part XIV.—*The species enumerated by D'Orbigny* in the "Annales des Sciences Naturelles," 1826, vol. vii. IV.—*The Species founded upon the Figures in Soldani's "Testaceographia ac Zoophytographia."* <*Ann. and Mag. Nat. Hist.*, ser. 4, vol. viii, pp. 145-179, 238-266. 1871.

PARKER, W. K., and T. R. JONES. On the Nomenclature of the Foraminifera. Part XV.—*The Species figured by Ehrenberg.* <*Ann. and Mag. Nat. Hist.*, ser. 4, vol. x, pp. 184-200, 253-271, 453-457. 1872.

PARKER, W. K., and T. R. JONES. On the Nomenclature of the Foraminifera. Part XV.—*The Species figured by Ehrenberg.* <*Ann. and Mag. Nat. Hist.*, ser. 4, vol. ix, pp. 211-230, 280-303. 1872.

PARKER, JONES and BRADY. On Priority in the Discovery of the Canal-system in Foraminifera. <*Ann. and Mag. Nat. Hist.*, ser. 4, vol. pp. 64 and 305. 1874.

PARKER, W. K., and T. R. JONES. On Ovulites margaritula. <*Ann. and Mag. Nat. Hist.*, ser. 4, vol. xx, p. 79. 1877.

PARKINSON, J. The Organic Remains of a former World. 3 vols. 4to. London, 1804-11.

PEACH, C. W. Additional List of Fossils from the Boulder-Clay of Caithness. <Report Brit. Assoc. (Bath Meeting) Trans. Soc., p. 61. 1864.

PEACH, C. W. Further Observations on, and additions to, the List of Fossils found in the Boulder-Clay of Caithness, N. B. Brit. Assoc. Advan. Sci., 1866, pp. 64, 65.
   See Brady.

PENNANT, T. The British Zoology. 8 vo. London, 1776–77.

PENNANT, T. British Zoology. London, 1812. "A new edition."

PERRY, G. Conchology or the Natural History of Shells. (Dentalia viridis, and bandata, pl. 52.) London, 1811.

PERRY, J. On collecting Foraminifera on the West Coast of Ireland. <Proc. Lit. Philos. Soc. Manches., vol. v, p. 42. 1866.

PHILLIPS, J. On the Remains of Microscopic Animals in the Rocks of Yorkshire. <Proc. Geol., and Polytech. Soc. W. R. Yorks., vol. ii, p. 277, pl. vii. Leeds. 1845.

PRESTWICH, J. On the Structure of the Strata between the London Clay and the Chalk in the London and Hampshire Tertiary Systems. Part III.—The Thanet Sands. <Quart. Journ. Geol. Soc. (Proc.), vol. viii, pp. 235–268, plates xv, xvi. 1852.

PRESTWICH, J. On the Thickness of the London Clay; on the Relative Position of the Fossiliferous Beds of Sheppy, Highgate, Harwich, Newnham, Bognor, etc. <Quart. Journ. Geol. Soc. Lond., vol. x, pp. 401–419. 1854.

PRESTWICH, J. On the Correlation of the Middle Eocene Tertiaries of England, France and Belgium. <Quart. Journ. Geol. Soc. Lond., vol. xii, pp. 390–392, 599–604. 1856.

PRESTWICH, J. On the Correlation of the Eocene Tertiaries of England, France and Belgium. <Quart. Journ. Geol. Soc. Lond., vol. xiii, pp. 89–134. 1857.

PRESTWICH, J. Notes on the Phenomena of the Quaternary Period in the Isle of Portland and around Weymouth. <Quart. Journ. Geol. Soc. Lond., vol. xxxi, pp. 29–52. 1875.

PRICE, F. G. H. A Monograph of the Gault, being the substance of a Lecture delivered in the Wooodwardian Museum, Cambridge, 1878, and before the Geological Association, 1879, p. 81. 1880.

PRITCHARD, A. A history of Infusoria, including the Desmidiaceæ and Diatomaceæ, British and Foreign, 4th edition, enlarged and revised by J. T. Arlidge, W. Archer, J. Ralfs, W. C. Williamson, and the Author. London, 1861.

PULTENEY, R. Catalogues of the Birds, Shells, and some of the most rare Plants of Dorsetshire, from the new and enlarged edition of Mr. Hutchin's History of that County. Fol. London, 1799.

READE, J. B. Observations on some new organic remains in the flint of chalk London, 1838.

READE, J. B. On the Animals of the Chalk still found in the living state in the Stomachs of Oysters. <*Trans. Micr. Soc. Lond.*, vol. ii, pp. 20-24. 1844.

REUSS, A. E., H. BRADY. Snyopsis of the Foraminifera of the Middle and Upper Lias, Somersetshire. <*Verhandl. K. K. Geol. Reich.* 1868, pp. 151, 152. 1868.

ROBERTSON, D. On Foraminifera from the South Coast of Devon and Cornwall. <*Report Brit. Assoc.*, (Exeter Meeting) p. 91. 1869.

ROBERTSON, D. Notes on the Recent Foraminifera and Ostracoda of the Firth of Clyde, with some Remarks on the Distribution of the Mollusca. <*Trans. Geol. Soc. Glasgow*, vol. v, p. 112. 1874.

ROBERTSON, D. Notes on a Raised Beach at Cumbrae. <*Trans. Geol. Soc. Glasgow*, vol. v, p. 192. 1875.

ROBERTSON, D. In G. S. Brady and Robertson's Report on Dredging off the Coast of Durham and North Yorkshire in 1874. <*Report Brit. Assoc.* (Bristol Meeting) p. 185. 1875.

ROBERTSON, D. Notes on a Post tertiary Deposit of Shell-bearing clay on the west side of the Railway Tunnel at Arkleston near Paisley. <*Trans. Geol Soc. Glasgow*, vol. v, p. 281. 1876.

ROBERTSON, D. Garnock-water Post-tertiary Deposit. <*Trans. Geol. Soc. Glasgow*, vol. v, p. 297. 1876.

ROBERTSON, D. Foraminifera in—A Contribution towards a Complete List of the Fauna and Flora of Clydesdale and the West of Scotland, p. 51, 8 vo. Glasgow, 1876.

ROBERTSON, D. Notes on the Post-tertiary Deposit of Misk Pit, near Kilwinning. <*Trans. Geol. Soc. Glasgow*, vol. v, p. 297. 1877.

ROBETSON, D. On the Post-tertiary Beds of gravel, Greenock. *Trans. Geol. Soc. Glasgow*, vol. vii, 1-37, pl. i. 1883.

ROBERTSON, D. Foraminifera, in D. J. Gwyn Jeffrey's paper,—Mediterranean Mollusca (No. 3.) and other Invertebrata. <*Ann. and Mag. Nat. Hist.*, ser 5, vol. xi, p. 401. 1883.

ROBERTSON, D. Report on the Sand and Gravels and Boulder-clays and the Silt, at the Dock F of the Atlantic Docks, Liverpool. (Appendix to T. Mellard Reade's paper,—The Drift beds of the North-west of England and North Wales.) <*Quart. Journ. Geol. Soc. Lond.*, vol. xxxix, pp. 129-132. 1883.

ROGERS, H. D. On the probable depth of the Ocean of the European Chalk Deposits. <*Proc. Bost. Soc. Nat. Hist.*, vol. iv, p. 297. 1853. *Amer. Journ. Sci.*, 2 ser., vol. xvii, p. 131. 1854.

SALTER, J. W. Arctic Carboniferous Fossils, collected by the Expedition under Sir E. Belcher, C. B., 1852-54, in the "last of Arctic Voyages," by Sir Edward Belcher. 2 vols., 8 vo. London, 1855; pp. 377, 391, pl. xxxvi.

SCHLUMBERGER, M. C. On Orbulina universa. <*Ann. Mag. Nat. Hist.*, ser. 5, vol. xiv, pp. 69-71. 1884.

SCHULTZE, M. S. Beobachtungen uber, die Fortpflanzung der Polythalamien. <*Muller's Archiv.*, 1856, p. 165. *Quart. Journ. Micr.*, vol. v, p. 220. 1856.

SEGUENZA, G. On Ellipsoidina, *a new Genus of Foraminifera*, with further Notes on its Structure and Affinities, by Henry B. Brady, F. L. S., F. G. S. <*Ann and Mag Nat. Hist.*, ser. 4, vol. i, pp. 333-343. 1868.

SHONE, W. On the Discovery of Foraminifera, etc. in the Boulder-clays of Cheshire. <*Quart. Journ. Geol. Soc. Lond.*, vol. xxx, pp. 181-185. 1874.

SHONE, W. On the Glacial Deposits of West Cheshire, together with Lists of the Fauna found in the Drift of Cheshire and the adjoining counties. <*Quart. Journ. Geol. Soc. Lond.*,, vol. xxxiv, p. 383; table. 1878.

SIDDALL, J. D. On the Foraminifera of the River Dee. <*Ann. and Mag. Nat. Hist.*, ser. 4, vol. xvii, p. 37. 1876.

SIDDALL, J. D. On Foraminifera and other Microzoa. <*Nature*, vol. xv, p. 461. 1878. (Abstract.)

SIDDALL, J. D. On the Foraminifera of the River Dee. <*Proc. Chester Soc. Nat. Sci.*, pt. ii. p. 42; wood cuts. 1878.

SIDDALL, J. D. Catalogue of British Recent Foraminifera, for use of Collectors. 8 vo. Chester. 1879.

SIDDALL, J. D. On Shepheardella, an Underscribed Type of Marine Rhizopoda, with a few Observations on Lieberkuehnia. <*Quart. Journ. Micr. Sci.*, vol. xx, n. s., p. 130, pls. xv, xvi. 1880.

SMITH, J. T. The Ventriculidæ of the Chalk, 8vo. London, 1848.

SOLLAS, W. J. An Aberrant Foraminifer. <*Nature*, vol. v, p. 83. Woodcut. 1871. *Peneroplis pertusus.*

SOLLAS, W. J. On the Foraminifera and Sponges of the Upper Greensand of Cambridge <*Geol. Mag.*, vol. x, pp. 268-274. 1873.

SOLLAS, W. J. On the Glauconite Granules of Cambridge Greensand. <*Geol. Mag.*, dec. II, vol. iii, p. 539, pl. xxi, 1876.

SOLLAS, W. J. On the Perforate Character of the Genus Webbina, with a notice of two new species, *W. laevis* and *W. tuberculata*, from the Greensand. *Geol. Mag.*, dec. II, vol. iv, p. 102, pl. vi. 1877.

SOLLAS, W. J. The Estuaries of the Severn and Its Tributaries; an inquiry into the nature and origin of their tidal sediment and alluvial flats. <*Quart. Journ Geol. Soc. Lond.*, vol. xxxix, pp. 611-626. 1883.

SOLLAS, W. J. On the Origin of Freshwater Faunas. A Study in Evolution. <*Sci. Trans. Roy. Dub. Soc.*, vol. iii, ser. II, pp. 87-118. 1884.

SORBY, H. C. On the Microscopical Structure of the Calcareous Grit of the Yorkshire Coast. <*Quart. Journ. Geol. Soc. Lond.*, vol. vii, pp. 1-6. 1851.

SORBY, H. C. Address delivered at the Anniversary Meeting of the Geological Society. <*Quart. Journ. Geol. Soc. Lond.*, vol. xxxv. with a privately published appendix of 18 plates. 1879.

SOWERBY, G. B. Foraminifera from the Colne Tidal River, Essex, 1 plate, (privately printed) 8vo. London, 1856.

SOWERBY, J. Mineral Conchology of Great Britain, 12 vols., 8vo. London, 1818-1829.

STEWART, S. A. A list of the Fossils of the Estuarine Clays of the Counties of Down and Antrim. <*Eighth Ann. Rept. Bel. Nat. F. C.*, appendix ii, pp. 27-40. 1871.

STEWARDSON, G., H. B. BRADY, and D. ROBERTSON. The Ostracoda and Foraminifera of Tidal Rivers, with an Analysis and Descriptions of the Foraminifera, by Henry B. Brady. <*Ann. and Mag. Nat. Hist.*, ser. 4, vol. vi, pp. 1-34, 273-309. 1870.

STRICKLAND, H. E. On two Species of Microscopic Shells found in the Lias. <*Quart. Journ. Geol. Soc. Lond.*, vol. ii, pp. 30, 31, wood cuts. 1846.

TATE, R. On the Correlation of the Cretaceous Formations of the north east of Ireland. <*Quart. Journ. Geol. Soc. Lond.*, vol. xxi, pp. 15-44, 3 plates. 1865.

TATE, R. and J. F. BLAKE. The Yorkshire Lias. 8vo., 19 plates and map. London, 1876.

THOMSON, C. W. The Depths of the Sea. 8vo. London, 1873.

THOMSON, C. W. The Depths of the Sea. Second edition, 8vo. London. 1874.

THOMSON, C. W. On Dredgings and Deep-Sea Soundings in the South Atlantic. <*Proc. Roy. Soc.*, vol. xxii, pp. 423-428. 1874.

THOMSON, W. On Deep Sea climates. <*Nature*, vol. ii, pp. 257-261. 1870.

THOMSON, W. The continuity of the chalk. <*Nature*, vol. iii, pp. 225-257-286. 1871.

THOMPSON, W. On the Fauna of Ireland. <*Mag. Nat. Hist.*, vol. v. 1840.

THOMPSON, W. Report on the Fauna of Ireland: Div. Invertebrata. <*Report Brit. Assoc.*, (Cork Meeting), p. 274. 1843.

THOMPSON, W. Additions to the Fauna of Ireland. <*Mag. Nat. Hist.*, vol. xiii. 1844.

THOMPSON, W. Report on the Fauna of Ireland. (Foraminifera) <*Brit. Assoc. Advan. Sci.* 1843, xiii, pp. 274, 275. 1844.

THOMPSON, W. "Preliminary Notes on the Nature of the Sea-bottom procured by the Soundings of H. M. S. 'Challenger' during her Cruise in the 'Southern Sea' in the early part of the year 1874." <*Proc. Roy. Soc.*, vol. xxiii, pp. 32-48. 1875.

THOMPSON, W. "Preliminary Report to the Hydrographer of the Admiralty on some of the Results of the Cruise of H. M. S. 'Challenger' between Hawaii and Valparaiso." <*Proc. Roy. Soc*, vol xxiv, pp. 463-470, 5 plates. 1876.

THORPE, C. British Marine Conchology; being a Descriptive Catalogue, arranged to the Lamarckian System, of the Salt-water Shells in Great Britain. 12mo. London, 1844.

TIZARD, Staff-Commander, and J. MURRAY. Exploration of the Faroe Channel during the summer of 1880, in Her Majesty's hired ship "Knight-Errant." <*Proc. Roy. Soc. Edinb.*, vol. xi, pp. 638-720, pl. vi,—Report on the Foraminifera by H. B. Brady, pp. 708-717. 1882.

TURTON, W. Linnæus. General system of Nature; translated from Gmelin's last edition, amended and enlarged. 8vo. Swansea, 1800-06.

TURTON, W. A Conchological Dictionary of the British Islands. 12mo. London, 1819.

TUTE, J. S. Organisms in Carboniferous Flint or Chert. *Science Gossip*, August 1874, p. 188. 1875.

VINE, G. R. Foraminifera from Shetland. *Science Gossip*, vol. xiv, p. 51. 1879.

VINE, G. R. Notes on the Carboniferous Entomostraca and Foraminifera of the North Yorkshire Shales. *Proc. Yorkshire Geol. and Pol. Soc.*, n. s., vol. viii, pp. 226-239. 1884.

WALFORD, E. A. On some Upper and Middle Lias Beds in the Neighbourhood of Banbury. <*Proc. Warwicksh. Field-Club* for 1878: Supplement. 1878.

WALKER, G. Testacea Minuta Rariora; a collection of the minute and rare shells lately discovered in the sand of the sea-shore near Sandwich, by William Boys, Esq. London, 1784.

WALLER, E. Report on the Foraminifera obtained in the Shetland Seas. <*Brit. Assoc. Advan. Sci.*, 1867, pp. 441-446. 1868.

WALLER, E. Report on the Shetland Foraminifera for 1868. <*Brit. Assoc. Advan. Sci.*, 1868, pp. 340, 341. 1869.

WALLICH, G. C. Notes on the Presence of Animal Life at Vast Depths in the Ocean. London, 1860. Privately printed, 8vo.

WALLICH, G. C. Remarks on some Novel Phases of Organic Life, and on the Boring Powers of Minute Annelids, at Great Depths in the Sea. <*Ann. and Mag. Nat. Hist.*, ser. 3, vol. viii, pp. 52-55. 1861.

WALLICH, G. C. The North Atlantic Sea-bed; comprising a Diary of the Voyage on Board H. M. S. "Bulldog," in 1860, and Observations on the Presence of Animal Life, and the Formation and Nature of Organic Deposits, at Great Depths in the Ocean, published with the Sanction of the Lords Commissioners of the Admiralty, part 1, with map and 6 pls. 4to. 1862.

WALLICH, G. C. On the mineral secretions of Rhizopods and Sponges. <Ann. and Mag. Nat. Hist., ser. 3, vol. xiii, p. 72.—Amer. Journ. Sci., vol. xxxviii, ser. 2, p. 131. 1864. A review.

WALLICH, G. C. On the process of Mineral Deposit in the Rhizopoda and Sponges, as affording a Distinctive Character. <Ann. and Mag. Nat. Hist., ser. 3, vol. xiii, pp. 72-82. Wood cuts. 1864

WALLICH, G. C. On the Deep-Sea Bed of the Atlantic and its inhabitants. <Quart Journ. Sci. Lond., vol. i, pp. 36-44. 1864.

WALLICH, G. C. On the extent and some of the principal causes of Structural Variation among the Difflugian Rhizopods. <Ann. and Mag. Nat. Hist., ser. 3, vol. xiii, p. 215, pls. xv., xvi. 1864.

WALLICH, G. C. On the Structure and Affinities of the Polycystina. <Trans. Micros. Soc. Lond., vol. xiii, p. 75-84. 1865.

WALLICH, G. C. On the Radiolaria as an Order of Protozoa. <Pop. Sci. Review, new series, vol. ii, pp. 267-368, pl. vi. 1868.

WALLICH, G. C. On some undescribed Testaceous Rhizopods from the North Atlantic Deposits. <Monthly Micr. Journ., vol. i, p. 104, pl. iii. 1869.

WALLICH, G. C. On the Vital Functions of Deep Sea Protozoa. <Month Micro. Journ., vol. i, p. 32. 1869.

WALLICH, G. C. On the Rhizopoda as embodying the Primordial Type Animal Life. <Monthly Micr. Journ., vol. i, p. 228. 1869.

WALLICH, G. C. On the true Nature of the so-called "Bathybius." <Ann. and Mag. Nat. Hist., ser. 4, vol. xvi, pp. 322-339. 1875.

WALLICH, G. C. Deep-Sea Researches on the Biology of Globigerina, 2 pls. 8vo. London, 1876.

WALLICH, G. C. On the Fundamental Error of constituting Gromia the Type of Foraminiferal Structure. <Ann. and Mag. Nat. Hist., ser. 4, vol. xix, p. 158. 1877.

WALLICH, G. C. Observations on the Coccosphere. <Ann. and Mag. Nat. Hist., ser. 4, vol. xix, p 342, pl. xvii 1877.

WALLICH, G. C. On Rupertia stabilis, a new sessile Foraminifer from the North Atlantic. <Ann. and Mag. Nat. Hist., ser. 4, vol. xix, p. 501, pl. xx 1877.

WALLICH, G. C. Deep Sea Researches on the Biology of the *Globigerina*. 1877.
   Not seen.

WALLICH, G. C. On the Radiolaria as an order of the Protozoa. <*Pop. Sci. Rev.*, new series, vol. vi, pp. 267-382, pl. vi. 1878.

WALLICH, G. C. A Contribution to the Physical History of the Cretaceous Flints. <*Quart. Journ. Geol. Soc. Lond.*, vol. xxxvi, pp. 68-92. 1880.

WALLICH, —. Note on the Detection of *Polycystina* with the hermetically closed Cavities of certain Nodular Flints. <*Ann. and Mag. Nat. Hist.*, ser. 5, vol. xii, pp. 52-53. 1883.

WALLICH, —. Critical Notes on Dr. Augustus Gruber's "Contributions to the Knowledge of the Amœbœ." <*Ann. and Mag. Nat. Hist.*, ser. 5, vol. xvi, pp. 215-227. 1885.

WEAVER, T. On the Composition of Chalk Rocks and Chalk Marl, from the Observations of Dr. Ehrenberg. <*Ann. and Mag. Nat. Hist.*, vol. vii, p. 398. 1841.

WETHERELL, N. T. Observations on a Well dug on the South side of Hampstead Heath. <*Trans. Geol. Soc. Lond.*, 2nd. ser., vol. v, p. 131, pl. ix. 1834.

WETHERELL, N. T. Notice of a species of *Rotalia* found attached to specimens of Vermetus Bognoriensis. <*Mag. of Nat. Hist.*, vol. iii, pp. 162, 163 1839.

WHITAKER, W. On the "Lower London Tertiaries" of Kent. <*Quart. Journ. Geol. Soc. Lond.*, vol. xxii, pp. 404-435. 1866.

WHITAKER, W. The Geology of the London Basin. <*Mem. Geol. Sur. Gt. Brit.*, vol. iv, pp. 575, 578, 581, 596, 600. 1872.
   Lists of Foraminifera found in Thanet, Woolwich and Reading, Oldhaven, London Clay, Bracklesham, and Upper Bagshot Beds.

WILSON, E. On the Occurrence of Foraminifera in the Carboniferous Limestone of Derbyshire. <*Midland Naturalist*, vol. iii, p. 220. 1880.

WILLIAMSON, W. C. On some of the Microscopical Objects found in the Mud of the Levant, and other Deposits, with remarks on the Formation of Calcareous and Infusorial Siliceous Rocks. <*Mem. Lit. and Philos. Soc. of Manchester*, ser. 2, vol. viii, p. 1. 1848.

WILLIAMSON, W. C. On the recent British species of the genus *Lagena* <*Ann., and Mag. Nat. Hist.*, ser. 2, vol. i, p. 1. 1848.

WILLIAMSON, W. C. On the Structure of the Shell and Soft Animal of *Polystomella crispa*, with some remarks on the Zoological position of the Foraminifera. <*Trans. Micros. Soc. Lond.*, vol. ii, p. 159, pl. xxviii. 1848.

WILLIAMSON, W. C. On the minute structure of the Calcareous Shells of some recent species of *Foraminifera*. <*Trans. Micros. Soc. Lond.*, ser. 2, vol. iii, p. 105. 1851.

WILLIAMSON, W. C. On the minute structure of a species of *Faujasina*. <*Trans. Micros. Soc. Lond.*, ser. 2, vol. i, p. 87. 1851.

WILLIAMSON, W. C. On the Recent Foraminifera of Great Britain. Printed for the Ray. Society, London, 1858.

WILLIAMSON, W. C. On the Anatomy and Physiology of the Foraminifera. <*Popular Science Review*, vol. iv, p. 171, pl. viii. 1865.

WILLIAMSON, W. C. Deep-sea Researches. <*Nature*, vol. xi, p. 148. 1875.

WILLIAMSON, W. C. On the Supposed Radiolarians and Diatoms of the Carboniferous Rocks. <*Report Brit. Assoc.* (Dublin Meeting), Trans. Sections, p. 534. 1878.

WILLIAMSON, W. C. The Origin of a Limestone Rock. <*Nature*, vol. xvii, p. 265. 1878.

WOOD, J. G. Common Objects of the Microscope, pp. 121, 122, n. d. 16mo.

WOOD, W. Index Testaceologicus; or a Catalogue of Shells, British and Foreign, arranged according to the Linnean System. 8vo. London, 1825.

WRIGHT, E. P. Fossil Calcareous Algæ. <*Nature*, xix, pp. 485, 486. 1879.

WRIGHT, J. A list of the Irish Liassic Foraminifera. <*Eighth Ann. Rept Bel. Nat. F. C.* 1870-71. Appendix ii, pp. 22-26. 1871.

WRIGHT, J. A list of the Cretaceous Microzoa of the North of Ireland. <*Proc. Bel. Nat. F. C.*, ser. ii, vol. i, appendix 1873-74, pp. 73-99. 1875.

WRIGHT, J. On the Discovery of Microzoa in the Chalk-flints of the North of Ireland. <*Rep. Brit. Assoc. Advan Sci.*, 1874, pp. 95, 96. 1875.

WRIGHT, J. Foraminifera, Recent and Fossil, with especial reference to those found in Ireland. <*Proc. Belfast Nat. Hist. and Phil. Soc.* Dec. 4, 1877.

WRIGHT. J. Recent Foraminifera of Down and Antrim. <*Proc. Belfast Nat. Field. Club*, 1876-7, appendix. 1877.

WRIGHT, J. Recent Foraminifera of Down and Antrim. <*Annual Rept. Bel. Nat. F. C.*, appendix iv, pp. 101-105, 1 plate, 2 folding lists. 1878.

WRIGHT, J. A list of the Post-Tertiary Foraminifera of the North-East of Ireland. <*Proc. Bel. Nat. F. C.*, appendix v, pp. 152-163. 1881.

WRIGHT, J. Notes on the Foraminifera, Genus Lagena. <*Proc. Bel. Nat. F. C.*, ses. 1880-81, pp. 108-109. 1882.

WRIGHT, J. A list of Recent Foraminifera found during the Belfast Naturalists' Field Club's Excursion to South Donegal, 1880. <*Proc. Bel. Nat. F. C.*, ser. ii, vol. ii, appendix vi, pp. 179-187, 1880-81, 1 plate. 1882.

WRIGHT, T. S. Description of New *Protozoa*. <*Edinb. New Philos. Journ.*, new series, vol. vii, pp. 276-281, 1858; vol. x, pp. 97-104. 1859.

WRIGHT, T. S. On the Reproductive Elements of the *Rhizopoda*. <*Ann. and Mag. Nat. Hist*, ser. 3, vol. vii, p. 360. 1861.

WRIGHT, T. S. Observations on British Protozoa and Zoophytes. <*Ann. and Mag. Nat. Hist.*, ser. 3, vol. viii, p. 120, pls. iii-v. 1861.

WYATT, J., and T. R JONES. On the *Orbitulinae* of the Chalk, and "Fossil Beds" of the Drift. Geol., p. 233. 1862.
    Not seen.

YOUNG, J., and J. ARMSTRONG. On the Carboniferous Fossils of the West of Scotland. <*Trans. Geol. Soc. Glasgow*, vol. iii, supplement. 1871.

YOUNG, J., and J. ARMSTRONG. The Fossils of the Carboniferous Strata of the West of Scotland. <*Trans. Geol. Soc. Glasgow*, vol. iv, pp. 267. 1873.

ALCOCK, T. On the Structure of the Shell of several common species of Polymorphina. <*Proc. Man. Lit., and Philo. Soc.*, vol. xxii, pp. 67, 68. 1883.

BRADY, H. B., in M'Intosh's Marine Invertebrates and Fishes of St. Andrews. (A list of Foraminifera,) pp. 11, 12. 1875.

CROSSKEY, H. W. Note on the *Ostracoda* and *Foraminifera* of the Shelly Patches at Bridlington Quay. <*Quart. Journ. Geol. Soc.*, vol. xl, pp. 325-327. 1884.

GARDNER, J. S. Chalk, and the Origin and Distribution of Deep-Sea Deposits. <*Nature*, vol. xxx, pp. 192, 193, 264, 265. 1884.

PART IV.

FRANCE AND ITALY.

FRANCE AND ITALY.

ACHIARDI, A. d. Corralli Fossili del Terreno *Nummulitico* dell' Alpi Venete. 4to. Pisa, 1867.

ANON. Foraminifères et Infusoires. Je donne ici l' énumération des espèces eu deux listes séparées ainsi qu elles ont été successivement publiées. <*Actes. Soc. Linn.*, ser. 3, vol. iv, pp. 643-651. 1861.

AOUST, VIRLET D'. Réponse aux différentes objections de *M. Viguier*, relatives à sa communication sur les *Marbres de l' Aude*. <*Bull. de la Soc. Geol. de France*, ser. 3, vol. xi, pp. 315, 318. 1883.

BACHMANN, I. Quelques remarques sur une note de M. Renevier intitulée: "Quelques observations géologiques sur les Alpes de la Suisse centrale. (Schwaytz, Uri, Unterwalden et Berne) comparées aux Alpesvaudoise." <*Mittheil der Naturforsch. Gesellsch. in Berne, Jahr*, 1869, pp. 161-174. 1870.

BARROIS, C. Recherches sur les terrains anciens des Astiories et de la Galicie. <*Mem. Soc. Geol. du Nord.*, vol. ii, pp. 1-630, pls. i-xx. (Also separately published, 1 vol., 4to, 20 plates), Lille. 1882.

BEAUMONT, É. de. Sur l' age du terrain nummulitique des Pyrénées. <*Bull. de la Soc. Geol. de France*, sér. 2, tome v, p. 413. 1848. (Leonhard's Jahr neuse buch für Geonosie, p. 272. 1848.

BECCARIUS, J. B. De Bononiensi arena quadam (*Commentarii de Bonon. Scient. et Art. Inst.*) Vol. i. 1731.

BELLARDI, L. Liste des fossiles de la formation nummulitique du comté de Nice. <*Bull. de la Soc. geol. de France*, sér. 2, vol. vii, pp. 678-683. 1850.

BELLARDI, L. Catalogue raisonné des Fossiles nummulitiques du comté de Nice. <*Mem. Geol. Soc. de France*, sér. 2, vol. iv, pp. 206-300, plates 12-22. 1852.

BELLARDI, L. Catalogo ragionato dei Fossili *Nummulitici* d' Equitto della raccolta del R. Museo Mineralogico di Torino. 4to. 1854.

BERTHELIN, G. Liste des Foraminifères receuillis dans la Baie de Bourgeneuf et à Pornichet. 8vo. Nantes, 1878.

BERTHELIN, G. Foraminifères du Lias Moyen de la Vendée. <*Revue et Mag. de Zool.*, 1879, p. 18, 1 pl. 1879.

BERTHELIN, G. Coup d' œil sur la Faune Rhizopodique du Calcaire Grossier inférieur de la Marne. <*Bull. de l' Assoc. France, pour l' Avance. des Sci.*, 1880, pp. 553-559. 1880.

BERTHELIN, G. Mémoire sur les Foraminifères fossiles de l'Étage Albien de Monteley (Doubs). <*Mem. Soc. Geol. de France*, sér. 3, vol. i, No. 5, pls. xxiv-xxvii. 1880.

BERTHELIN, G. Les Foraminifères Fossiles de l' étage Albien de Monteley, 4 pls. Paris, 1882.

BERTHELIN, G. Sur l'ouverture de la Placentula Partschiana, d'Orb., sp. <*Bull. de la Soc. Geol. de France*, sér. 3, vol. xi, Nov. 6th, pp. 66, 17. 1882.

BERTHELIN, G. Réponse à la *Note de M. Terquem*, au sujet de l'ouverture de la *Placentula Partschiana*. <*Bull. d la Soc. Geol. de France*, sér. 3, vol. xi, pp. 304-308. 1882.

BERTHELIN, G. Liste des Foraminifères recueillis dans la baie de Bourgeneuf et à Pornichet. Nantes, 55 pp., 8vo. 1884.
Not seen.

BEUDANT, F. S. Cours Élémentaire d'Histoire Naturelle. La Minéralogie et la Géologie, pp. 116-118, 5th edition. Paris, 1851.

BEUDANT, F. S. Cours Élémentaire d'Histoire Naturelle. La Minéralogie et la Géologie. Calcaires a *Nummulites*, pp. 239-240, 5 edit. Paris, 1851.

BLAINVILLE, H. M. Ducrotay de, Traité de Malacologie. Paris, 1825.

BLAINVILLE, H. M. *Ducrotay de*, Manuel de Malacologie et de Conchyliologie, &c. Paris, 1825-27.

BLAINVILLE, H. D. de. Faune Française, Malacazoaires on Animaux Mollusques. Paris, 1820-30.

BLAINVILLE, H. D. de. Manuel de l'Actinologie ou de Zoophytologie. 8vo. Paris, 1834.

BLAINVILLE, H. D. de. Dictionaire des Sciences Naturelles. Paris, 1814-30.

BOEHM, G. Contribuzione allo studio dei calcari grigi del Veneto. <*Boll. d. R. Com. Geol. d.' Italia*, ser. ii, vol. vi, pp. 156-165. 1885. (G. B. C.)

BONISSENT,— Essai Géologique sur le Départment de la Manche, 9e Époque. —Sol Secondaire. Terrani Crétacé. <*Mem. Soc. Sci. Nat. Cher.*, vol. xi, pp. 217-228. 1865.

BONISSENT, — Essai Géologique sur le Départment de la Manche, 10e Epoque.—Sol Tertiaire. <*Mem. Soc. Sci. Nat. Cher.*, vol. xiii, pp. 5-34. 1867.

BORNEMANN, L. G Sopra una Specie mediterranea del genere Lingullnopsis. <*Atti della Soc. Tosc. Sci. Nat.*, vol. vi, fasc. 1, and plate. 1883.

BOSC, L. A. G. Histoire Naturelle des Coquilles. Paris, 1802.

BOUE, A. Observations sur le travail de M. Adolphe de Morlot relatif à la position du calcaire à Nummulites relativement, au grès à Fucoides de Vienne et de Trieste et au calcaire crétacé à Rudistes. <*Bull. de la Soc. Geol. de France*, sér. 2, vol. v, p. 68. 1847.

BOUBEE, N. Observations sur la note de M. d'Archiac relative aux fossiles du terrainé a nummulites de Bayonne et de Dax. *Bull. de la Soc. Geol. de France*, sér. 2, vol. iv, pp. 10, 11. 1847.

BRUGUIERE, J. G. Encyclopédie Méthodique. *Hist. Nat. desiere*, vol. i. Paris, 1789.

Duvignier, A. Statistique Géologique, minéralogique, metallurgique et paléontologique du départment de la Meuse. 8vo and 32 plates 4to. Paris, 1852.

Caillaux, A. Sur le terrain nummulitique en Toscane. <*Bull. de la Soc. Geol. de France.*, sér 2, vol. viii, pp. 131-136. 1851.

Cailliaud, F. Voyage à Méroé, au Fleuve blanc, etc. 4 vols. Paris, 1827.

Capellini, G. Calcare a Amphistegina, strati a Congeria e calcare di Leitha dei Monti Livornesi, nuove. <*Boll. R. Comit. Geol. D' Italia*, vol. vi, pp. 241-244. 1875.

Cattaneo, G. Prime ricerche sui Protozoi, 12 pp. Pavia, 1878.

Catullo, A. Sur l'inadmissibilité de la Faune fossile annoncée par M. Ewald comme caracteristique de la grande formation nummulitique du terrain tertiaire, 12 pp. Padone, 1848.

Claparede, Edouard, et Lachmann. Etudes sur les *Infusoires* et les *Rhizopodes*. Genève, 1858-61.

Coppi, F. Frammenti di Paleontologia Modense. <*Boll. del. R. Com. Geol. Anno.*, 1876, No. 5-6. 1876.

Coppi, F. Sul calcare Zancleano? Estratto dagli. <*Atti. Soc. dei. Nat. di Modena.*, ser. iii, vol. i. 1883.

Coppi, F. Il. Miocene medio nei colli modenese; appendice alla Paleontologia Modenese. <*Boll. R. Comit. Geol. D'Italia.*, vol. xiv, pp. 171-201. 1884.

Cornuel, M. J. Description des nouveaux fossils microscopiques du terrain crétacé inférieur du départment de la Haute-Marne. <*Mem. Geol. Soc. de France*, sér. 2, vol. iii, pp. 241-263, 2 plates. 1848.

Cornuel, J. Catalogue des coquilles de mollusques entomostracés et foraminifères du terrain crétacé inferieur de la Haute-Marne, avec diverses observations relatives à ce terrain. <*Bull. de la Soc. Geol. de France*, sér. 2, vol. viii, pp. 430-448. 1851.

Costa, O. G. Fauna del Regno di Napoli. Naples, 1838.

Costa, O. G. Foraminiferi Fossili della Marna Blu del Vaticano. <*Mem. Accad. Sci. Napoli*, vol. ii, p. 113, pl. i. 1855.

Costa, O. G. Foraminiferi Fossili delle Marne Terziarie di Messina. <*Mem. Accad. Sci. Napoli*, vol. ii, p. 127, pls. i, ii,—continuazione, ibid, p. 367, pl. iii. 1855.

Costa, O. G. Paleontologia del Regno di Naopli, parte 2. <*Atti. dell' Accademia Pontaniana*, vol. vii, p. 105, pls. ix-xxvii. 1856.

Costa, O. G. Microdoride Mediterranea, o Descrizione de' poco ben conosciuti od affatto ignoti viventi minuti e microscopici del Mediterraneo, vol. i. Naples, 1861.

Costa, O. G. Sopra i foraminiferi di Messina e della calabria estrema. <*Rendic. dell' Accad. d. sci. fis. e matem. di Napoli*, vol. v, pp. 366-372. 1866.

Cuvier, Geo. L. C. F. Le Règne Animal distribué d'après son Organisation. Paris, 1817.

Cuvier, Geo. Le Règne Animal, distribué d'après son Organisation, pour servir de base à l'histoire naturelle des animaux et d'introduction à l'anatomie comparée. 2nde, edit. Paris, 1828-30.

D'Archiac, Le Vicomte. Mémoire sur la formation crétacée du Sud-Ouest de la France. <*Mem. Soc. Geol. de France*, vol. ii, pp. 157-192. 1835.

D'Archiac. Mémoire sur la formation crétacée du sudouest de l France. <*Mem. Soc. Geol. de France*. 1837.

D'Archiac, L. V. Essais sur la coordination des terrains tertiaires du nord de la France et de l'Angleterre. <*Bull. Soc. Geol. de France*, vol x, p. 168. 1839.

D'Archiac, L. V. Sur les caractères tirés de la différence de stratification et le classement des terrains à nummulites <*Bull de la Soc. Geol. de France*, sér. 1, vol. iv, pp. 532-536. 1843.

D'Archiac, L. V. Observations sur divers terrains à nummulites et sur leur classement. <*Bull de la Soc. Geol. de France*, sér. 1, vol. iv, pp. 485-491. 1843.

D'Archiac, L. V. Description des fossiles recueillis par M. Thorent aux environs de Bayonne (extrait). <*Bull. de la Soc. de Geol. France*, sér. 2, vol. iii, pp. 475-477. 1846.

D'Archiac, L. V. Description des fossiles recueillis par M. Thorent dans les couches à Nummulines des environs de Bayonne. <*Mem. Geol. Soc. de France*, sér. 2, vol. ii, pp. 189-217, plate 7. 1846.
Calcarina? stellata, Nov. sp.

D'Archiac. Sur les fossiles à Nummulites des environs de Bayonne et de Dax. <*Bull. de la Soc Geol. de France*, sér. 2, vol. iv, pp. 1006-1013. 1847.

D'Archiac, A. Déscription des Fossiles du groupe Nummulitique recueillis par M. S.-P. Pratt et M J. Delbos aux environs de Bayonne et de Dax. <*Mem. de la Soc. Geol. de France*, sér. 2, vol. iii, pp. 397-456, pls viii-xiii. 1848.

D'Archiac. Historie des progrès de la Géologie de 1834-59, 8 vols., vol. iii. Paris, 1850.

D'Archiac, A. Description des fossiles du groupe nummulitique, recueillis par M. M S —P. Pratt et J. Delbos aux environs de Bayonne et de Dax. *Mem. Geol. Soc. de France*, ser. 2, vol. iii, pp. 397-502, plates, 8, 9. 1850.

D'Archiac, Le. V., in Bellardi's Catalogue raisonné des Fossiles Nummulitiques du Comté de Nice. <Mem. Soc. Geol. France, sér. 2, vol. iv, p. 204, pls. xiv-xxii. 1852.

D'Archiac, Le. V. Description de quelques fossiles nouveaux ou imparfaitement connus des environs des Bains de Rennes. <Bull. Soc. Geol. de France, sér. 2, vol. xi, p. 205, pl. ii. 1854.

D'Archiac, L. V. Observations critiques sur la distribution stratigraphie et synonymie de quelques rhizopodes. <Bull. Soc. Geol. de France, sér. 2, vol. xviii, pp. 460-468. 1861.

D'Archiac, M. Etudes Géologiques d'une partie des départements de l'Aude et des Pyrénées—Orientales. <Mem. Geol. Soc. de France, 2 sér., vol. vi, (Groupe Nummulitque) pp. 288-311. 1859.

D'Archaic et Jules Haime. Description des animaux fossiles du groupe nummulitique de l'Inde (extrait). <Bull. de la Soc. Geol. de France, sér. 2, vol. x, pp, 378-384. 1853.

D'Allard, De Sarran. Recherches sur les Dépôts fluvis-lacustres antérieurs et postérieurs aux assises marines de la craie supérieure du départment du Gard. <Bull. Soc. Geol. de France, sér. 3, vol. xii, pp. 553-634. 1884.

De Cristofori, J. Conchylia Fossileo exformatione telluris tertiaria in collecti nostra extrantia. 1832.
    Not seen.

De Favanne. La Conchyliologie, ou histoire naturelle des Coquilles de mer, d'eau douce, terrestres et fossiles, etc ; Dezallier d'Argenville, augentée par De Favanne de Mentcervelle, père et fils. Paris. 1786.

De Folin, Le Marquis. Exploration du Travilleur, 1880—Golfe du Gascogne, Rhizopodes Réticulaires, liste des genres et espèces. <Bull. Soc. d'Hist. Nat. de Toulouse, pp. 12. 1881.

Defrance, J. L. M. Art.—Nummulites, etc. <Dictionnaire des Sciences Naturelles, vol. xxxv. 1825.

De Grateleup, J. P. S. Catalogue Zoologique renfermant les débris fossiles des Animaux. etc., dubassin de la Gironde. 8vo, Bordeaux, 1838.

De la Harpe, P. Note sur les Nummulites des Environs de Nice et de Menton. Lettre à M. le Prof. Renevier, par M. Phil. de la Harpe. <Bull. Soc. Geol. de France, sér. 3, vol. v, p. 817. 1877.

De la Harpe, P. Note sur les Nummulites des Alpes occidentales. <Actes de la Soc. Helvet Sci. Nat. S. 60; pp. 225-233. 1878.

De la Harpe, P. Les Nummulites du Comté de Nice. leur espèces et leur distribution stratigraphique, et Échelle des Nummulites. <Bull. Soc. Vaud. Sci. Nat., vol. xvi, pp. 201-243; 1 plate. 1879.

De la Harpe, P. Nummulites des Alpes Françaises. <Bull. Soc. Vaud. Sci. Nat., vol. xvi, pp. 409-434. 1879.

DE LA HARPE, P. Description des Nummulites appartenant à la Zone supérieure des Falaises de Biarritz. <*Bull. Soc. de Borda. a Dax*, IV année. 1879.

DE LA HARPE, P. Description des Nummulites des falaises de Biarritz; additions et conclusions. <*Bull. Soc. de Borda. a Dax*, vol. vi. 1881.

DE LA HARPE, P. Note sur la distribution par couples des Nummulites Éocènes. <*Bull. Soc. Vaud. Sci. Nat.*, vol. xvii, pp. 429-444. 1881.

DE LA HARPE, P. Sur l'importance de la loge centrale chez les Nummulites —lettre de *M*, de la Harpe. <*Bull. Soc Geol. France*, vol. ix, pp. 171-176. Tournouer, M. Observations, ibid., pp. 176-178. 1881.

DE LA HARPE, P. Etud des Nummulties de la Suisse et révision des espèces éocènes des genres Nummulties et Assilina, pt. I. <*Mem. Soc. Paleont. Suisse*, vol. vii, pp. 1-104, pl. i, ii. Pt. II, ibid., pp. 105-140. 1881.

DELBOS, J. Remarques sur les la note de *M*. d'Archiac, relative aux fossiles du terrain à nummulites de Bayonne et de Dax. <*Bull. de la Soc. Geol. de France*, sér. 2, vol. iv, pp. 1013. 1847.

DELBOS, J. Notice sur les fahluns du Sud-Ouest de la France. <*Bull. de la Soc. Geol. de France*, sér. 2, vol. v, pp. 417-428. 1848.

DELUC, G. A. Observations sur la Bélemnite. <*Journ. de Physique de Chimie. et d'Hist. Nat.*, vol. lii, pp. 362-366. 1801.

DELUC, G. A. La Lenticulaire Numismale et la Bélemnite. *Journ. de Physique de Chimie et, d'Hist. Nat.*, vol. liv, pp. 173-180, 1 plate. 1801.

DELUC, G. A. Nouvelles observations sur la Lenticulaire de la Perte du Rhone et la Lenticulaire Numismale. *Journ. de Physique de Chimie. d'Hist. Nat.*, vol. lvi, pp. 325-346, 1 plate. 1803.

DESHAYES, G. P. Description des Coquilles fossiles des environs de Paris. 1824.

DESHAYES, G. P. Mémoire sur les *Alveolines*, etc. <*Annales des Sciences Naturelles*, vol. xiv, p. 225. 1828.

DESHAYES, G. P. Encyclopédie Méthodique; Histoire naturelle des Vers, des Mollusques, des Coquellages, Zoophytes; Hist. Nat. des Vers, par Bruguière et de Lamarck, continuée by G. P. Deshayes. Paris. 1830-32.

DESHAYES, G. P. Description des Coquilles caracteristiques des Terrains. Strasbourg. 1831.

DESHAYES, G. P. Observations sur les Fossiles de la Crimée. Mém. Géol. sur la Crimée par M. de Verneuil. *Mem. Soc. Geol. de France*, vol. iii, p. 1, pls. i-vi. 1837.

DESHAYES, G. P., and MILNE EDWARDS, in Lamarck's Animaux sans Vertèbres. 2de ed., vol. xi. 1846.

DESLONGCHAMPS, E. Zoophytes (Encyclopédie Méthodique). 4to. Paris. 1824.

DE STEFANI, C. Quadro comprensivo dei Terreni che costituiscono l'Apennino settentrionale. <*Atti della Soc. Tosc. Sci. Nat.*, vol. v, pp. 206-253. 1881.

DILLWYN, L. W. A Descriptive Catalogue of recent Shells, arranged according to the Linnean method, with particular attention to synonymy. London. 1817. Discussion sur les terrains nummulitiques du midi de la France. <*Bull. de la Soc. Geol. de France*, sér. 2, vol. iv, pp. 537-542, 560. 1847.

DODERLEIN, P. Cenni geologici intorno alla giacitura dei terreni miocenici superiori dell' Italia centrale. <*Atti dei Scieneifici Italiani*, vol. x, pp.— Sienna, 1862.

D'ORBIGNY, ALCIDE Dessalines. Modèles de Céphalopodes Microscopiques vivans et fossiles, représentant un individu de chacun des genres et des sous-genres de ces Coquilles, etc. Paris, 1826; 2nd edit., Paris, 1843.

D'ORBIGNY, M. D. Tableau méthodique de la classe des Céphalopodes. Ordre.—Foraminifères. <*Ann. d. Sci. Nat.*, vol. vii, pp. 245-314, plates x-xvii. 1826.

D'ORBIGNY, A. Sur les Foraminifères de la Craie Blanche de Paris. <*Mem. de la Soc. Geol. de France*, vol. iv, p. 1, pls. iv. 1839.

D'ORBIGNY, M. A. Mémoire sur les Foraminifères de la Craie Blanche du Bassin de Paris. <*Mem. Soc. Geol. de France*, 1 sér., vol. iv, pp. 1-51, plates 1-4. 1840.

D'ORBIGNY, A. Dictionnaire Universelle d'Histoire Naturelle, vol. v, p. 662-671. Art. Foraminifèras. Paris, 1844.

D'ORBIGNY, A. Prodrome de Paléontologie stratigraphique universelle des Animaux Mollusques et Rayonnés, 3 vols., 8o, with 4to Atlas. Paris, 1849-52.

D'ORBIGNY, A. D. Cours élémentaire de Paléontologie et de Géologie stratigraphiques, vol. ii, fasc. 1,800, and Atlas 4to. 1852.

DUJARDIN, F. Observations nouvelles sur les prétendus Céphalopodes microscopiques. <*Ann. de la Societe des Sciences nat. de France*, sér. 2, vol iii, pp. 108, 312. 1835.

DUJARDIN, F. Recherches sur les organismes inférieurs. <*Annales des Sciences naturelles*, sér. 2, vol. iv, p. 343. 1835.

DUJARDIN, F. Observations sur les Rhizopodes et les Infusoires. <*Comptes Rendus*, 1835, p. 338. L'Institut. 1835, No. 11, p 202. 1835.

DUJARDIN, F. Histoire Naturelle des Zoophytes, Infusoires, comprenant la physiologie et la classification de ces animaux et la maniere de les étudier à l'aide du microscope. 22 plates. 1841.

DUJARDIN. Notice sur les Infusoires. 1845.

DUJARDIN, F. Art.-Infusoires. <*Dictionnaire universel d'Histoire naturelle*, vol. vii, pp. 43-52, 1848. Art.-Rhizopodes, Ibid., vol. xi, pp. 115, 116. 1846.

DUJARDIN. Articles: Orthocère; in 8, 4 colonnes, Orucule: in-8, 1 colonne. Opis, Pachymia, Pentacrinites, Pentremites, Operculina, Orbitolite, Ovulite; in-8, 1 colonne. <*Dictionnaire universel d'histoire nat.* rédigè par M. C. d'Orbigny. 1847.

DUTHIERS, H. LACAZE. Mémoire sur les Antipathaires. (*Genre Gerardia*, L. D.) <*Ann. des Sci. Nat.*, sér. v, vol. ii, pp. 169-239. 1864.

EWALD, J. Remarques sur les Nummulites. Padoue, 1847.

EWALD, J. Quelques remarques sur les nummulites, 7 pp. Padoue, 1848.

FAUVERGE, H. G. Sur le dépôt à Nummulites du département de l'Aude. *Bull. de la Soc. Geol. de France*, sér. 2, vol vii, pp. 633-636. 1850.

FERRY, H. DE. Mémoire sur le groupe oolithique des inférieur des environs de Mâcon (Saône-et-Loire). 1 re partie. Jura mâconnais, étage bajocien. 4o laen.

FERUSSAC, LE BARON and E I. D'AUDEBARD d. Tableaux Systématiques des Animaux Mollusques, etc. Paris and London, 1822.

FERUSSAC, M. De. Bulletin universel des Sciences et de l'industrie. Paris, 1827.

FERUSAC, —. Aperçu historique sur les Céphalopodes. 1835.

FISCHER, P. Bryozoaires, Echinodermes, et Foraminifères marins du Département de la Gironde et des Côtes du Sud-Ouest de la France. *Actes de la Soc. Linn de Bordeaux*, vol. xxvii, (Foraminifera), p. 377. 1870.

FISCHER, P. Examen d'une série de sondages éxecutés dans l'atlantique sous las direction du commadant Vignes. *Journ. de Zool*, vol. iv, pp. 298-302. 1875.

FISCHER, P. Note sur un type particulièr de Rhizopodes. (Astrorhiza) *Journ. de Zool.*, vol. iv, pp. 503-510, 1 plate. 1875

FONTANNES, F. Première Note sur les Foraminifèrs des Terrains tertiaires supérieurs du Bassin du Rhône. *Ann. Soc d'Agric. Hist. Nat , et arts utiles de Lyon*, ser. 5, vol. ii, pp. 199-203. 1880.

FORBES, E. Note on the Fossils found by Lieut. Spratt in the several beds of the Teritary Formation of Malta and Gozo. *The Geology of Malta and Gozo.*, pp. 21-24. 1852.

FORNASINI, C. Nota preliminare sui Foraminiferi della marna pliocenica del Ponticello di Savena nel Bolognese. *Boll. Soc. Geol. Ital*, vol. ii, (16 pages, 1 plate). 1883.

FORNASINI, C. I foraminiferi della Tabella Oryctographica esistente nel R. Museo Geologico di Bologna. *Boll. d. R. Com. Geol. d' Italia*, ser. ii, vol. vi, pp. 53, 54. 1885. (M. C.)

FORTIS, C. A. Quelques novelles espèces de Discolithes. (Camerines, Lenticulaires, Helicites, Numismales, etc., (A letter to C. Hermann). <*Journ. de Physique de Chimie et d'Hist. Nat*., vol. lii, pp. 106-115, 1 plate. 1801.

FORTIS, J. B. Mémoires pour Servir à l'Histoire naturelle, et principalement a l'oryctographie del'Italie, 2 vols. Paris, 1802.

FUCHS, TH , and A. MANZONI. Sulla Relazione un viaggio Geologico in Italia. <*Boll. del. R. Com. Geol. d'Italia*, 1875, No. 7 e 8. Nota di G. Seguenza. Ibid., No. 9 e 10. 1874.

GAUDIN, C. T., and M. MOGGRIDGE. Mémoires. Menton. I. Terrains secondaires Crétacé. II. Eocène, Pliocène. <*Bull. Soc. Vaud. d. Sci. Nat.*, vol. viii, pp. 187-197. 1865.

GEMMELLARO, G. G. Sopra taluni organici Fossili del Turoniano e Nummulitico di Judica. <*Atti. dell. Accad. Gioenia di Sci. Nat.*, ser. 2, vol. xv, p. 269. 1860.

GERVAIS, P. Sur un Point de la Physiologie des Foraminifères. <*Comptes Rendus*, p. 469, 1847, and in L'Institut, p. 316. 1847.

GINANNI, G. CONTE. Opere postume nelle quale si contengono 114 Piante che Vegetano nel mare Adriatico da lui osservante, e descritte. Fol. Venezia, 1755-57.

Gregorio, A. de. Sulla Fauna delle argille Scagliose di Sicilia (olgiocene-eocene) e sul Miocene di Nicosia. Palermo, 1881. Tav. III, Fig. 3 a. b. *Alveolina Sphaeroidea* (Fort).

Gregorio, A. de. Fossili dei Dintorni di Pachino, 22 pp., 6 plates. Palermo, 1882.

GUALTIERI, N. Index Testarum Conchyliorum quæ adservantur in museo Nic. Gualtieri, et methodice distributæ; exhibentur; tabulüs œn. cx, Fol. maj. Florentinæ. 1741.

GUETTARD, J. E. Nouvelle Collection de Mémoires sur différentes parties intéressantes des Sciences et des Arts 3 vols. 4to. Paris, 1786.

GUMBEL, C. W. Beitrage zur Foraminiferenfauna der Nordalpinen Eocangebilde. <*Abh. d. II. cl. d. k. Ak. d. Wiss.*, Bd. x, II, abth. pp. 581-730. 1868.

GUMBEL, C. W. Geognostische Mittheilungen aus den Alpen. <*Sitzung. d. Math. phys. classe d. k. b, Akad. Munchen*, Bd. iii, pp. 37-40. 1873.

GUMBEL, C. W. Kurze Anleitung zu geologischen Beobachtungen in den Alpen. <*Anleit. zu Wissen. Beobacht. auf alpenreisen*, pt. i, p. 25; wood cuts. 12mo. Munich, 1878.

GUMBEL, C. W. Geognostische Mittheilungen aus den Alpen. <*Sitzungsber d. Math-phys. classe d. k. b. Akad. Wiss. Munchen*, 1880; pp. 164-240. 1880.

HAMILTON, W. J. Sulla Costituzione Geologica dei Monti Pisani. Memoria del Prof. Cav. Paolo Savi, Pisa, Presso rocco Vannucchi, 1846. <*Quart Journ. Geol. Soc. Lond*, vol. iii, part ii, pp. 1-10. 1847.

HERBERT, E. Note sur le terrain nummulitique de' l' Italie septentrionale et des Alpes, et sur éoligocène d'Allemagne. <*Bull. d. la Soc. Geol. de France*, ser. 2, vol. xxiii, pp. 126-144. 1865.

HERBERT, E. Sur le *Groupe nummulitique* du *Midi* de la France. <*Bull. de la Soc. Geol. de France*, ser. 3, vol. x, pp. 364-392. 1882.

HERBERT, E. Aperçu général sur la Geologie des environs de Foix. <*Bull. Soc. Geol. de France*, sér. 3, vol. x, pp, 523-531. 1882.

HERBERT, E. Disposition du terrain tertiaire à Lavelanet. <*Bull Soc. Geol. de France*, sér. 3, vol. x, pp. 565-569. 1882.

HERBERT, E. Sur la structure géologique du vallon de Paradières. *Bull. Soc. Geol. de France*, sér. 3, vol. x, pp. 548-551. 1882.

HERBERT, E. Compte rendu de l'excursion de St.-Girons à Ste Croix. <*Bull. Soc. Geol. de France*, sér. 3, vol x, pp. 615-622. 2 wood-cuts. 1882.

HERBERT, E. Compte rendu de l'excursion de de Sainte Croix à Audinac. <*Bull. Soc. Geol. de France*, sér. 3, vol. x, pp. 623-631. 1882.

HERBERT, E. Allocution finale et résumé des observations faites par la Société pendant la session de Foix. <*Bull. Soc Geol. de France*, sér. 3, vol. x, pp. 644-659. 1882.

HERBERT, E. Sur la faune de l'etage danien dans les Pyrénées. <*Bull. Soc. Geol. de France*, sér. 3, vol. x, pp. 664-666. 1882.

HEBERT, E. et RENEVIER. Description des fossiles du terrain nummulitique de Gass, des Diablerets, etc. Extrait du Bulletin de la Soc de Statistique du Dép. de l'Isere, Grenoble. 1854.
   Title not seen; taken from foot note.

ISSEL, A. Esame sommario di alcuni saggi di fondo raccolti nel Golfo di Genova. <*Boll. d, R. Com. Geol. d. Itali z.*, ser. ii, vol. vi, pp. 129-148. 1885.

JOLY, N. and LEYMERIE. Principaux résultats de leur recherches sur les Nummulites. <*Comptes Rendus*, vol. xxv, p. 591. 1847.

JOLY, N. et A. LEYMERIE. Mémoire sur les Nummulites considérées Zoologiquement et geologiquement. <*Mem. de l'Acad. des. Sci. de Toulouse*, vol. iv, p. 149, pl. i. 1849.

JONES. T. R On the Fossil Foraminifera of Malta and Gozo. <*Geologist*, vol. vii, pp. 133-135. 1864.

JONES, T. R., and W. K. PARKER. Appendix to Ansted Geology of Malaga. <*Quart. Journ. Geol. Soc. Lond.*, vol. xv, pp. 585-604. 1859.

LAMARCK, J. B. P. de. Mémoire sur les fossiles environs de Paris, comprenant la détermination des espèces qui appartiennent aux animaux marins sans vertèbres. 4to. Paris, 1802-06.

LAMARCK, J. B. de. Suite des Mémoires sur les Coquilles fossiles des environs de Paris. <Ann. du Mus., vol. v, 1804; vol. vii, 1806; vol. ix, 1807.

LAMARCK, J. B. P. Ant. De MONNET. Extrait du cours de Zoologie du Muséum d'Histoire naturelle sur les animaux invertébrés. Paris. 1812.

LAMARCK, J. B. de. Tableau Encyclopédique et Méthodique des Trios Régnes de la Nature; vingt-troisième partie; mollusques et Polypes divers. Paris. 1816.

LAMARCK, DE LA. Recueil de Planches des Coquilles Fossiles des Environs de Paris. 38 plates. 4to. Paris. 1823.

LAMARCK, J. B. P. Ant. DE MONNET. Système des animaux sans vertèbres, ou tableau général des classes, des ordres et des genres des ces animaux. Paris 1801.

LAMARCK, J. B. P. Ant. De MONNET. Mémoire sur les fossiles des environs de Paris, comprenant la détermination des espèces qui appartiennent aux animaux marins sans vertèbres. Paris. 1804.

LAMARCK, J. B. P. A. de. Histoire naturelle des Animaux sans vertèbres, vol. ii, 1816; vol. vii, 1822, 2nd edit. 1835-45.

LAMARCK, J. B. de. Histoire naturelle des Animaux sans vertèbres; 1 ère édit., Paris, 1815-1822; 2 ième édit., augmentée de notes par M.M. Deshayes et Milne-Edwards, Paris, 1835-1842.

LATRIELLE, PIERRE ANDRE. Familles naturelles du règne Animal, etc. Second edition. Paris. 1825.

LEYMERIE et JOLY, N. Mémoire sur les nummulites considérées zoologiquement et géologiquement. Voyez Joly et Leymeric.

LEYMERIE, A. Note sur le terrain nummulitique de la Scile et considérations général à ce sujet. <Bull. de la Soc. Geol. de France, ser. 2, vol. ii, pp. 27. 1844.

LEYMERIE, A. Litle résumé d'un Mémoire sur le terrain à Nummulites (épicrétacé) des Corbières et de la Montagne Noire (Aude). <Bull. de la Soc. Geol. de France, sér 2, vol. ii, pp. 11-27. 1844.

LEYMERIE, A. Lettre sur le terrain à Nummulites des Corbières. <Bull. de la Soc. Geol. de France, sér 2, vol ii, pp. 270-273. 1844.

LYMERIE, M. A. Mémoire sur le terrain à nummulites (épicrétacé) des Corbières et de la Montagne Noire. <Mem. Soc. Geol. de France, sér. 2, vol. i, pp. 327-373, and plate 13. 1845.

    Nummulites Atacicus, Leym. N. globulus, Leym. Operculina ammonea, Leym. O. granulosa, Leym. f Alveolina sub-Pyrenaica, Leym. A. var. globosa, Leym.

LEYMERIE, A. Mémoire sur le terrain à Nummulites (épicrétacé) des Corbières de la Montagne noire. <Mem. Soc. Geol. de France, ser. 2, vol. i, p. 337, pls. xii-xvii. 1846.

LEYMERIE, A. Observations critique (1) sur une note de M. Raulin, intitulée. Quelques mots encore sur le terrain à Nummulites. <*Bull. de la Soc. Geol. de France*, sér. 2, vol. vii, pp. 90-98. 1850.

LEYMERIE. Mémoire sur un nouveau type pyrénéen. <*Mem. Societe Geol. de France*, vol. iv. 1851.

LEYMERIE, A. Observations sur quelques terrains de la Provence. <*Bull. de la Soc. Geol. de France*, ser 2, vol. viii, pp. 202-207. 1851.

LEYMERIE, A. Note sur quelques localités de l'Aude, et particulièrement sur certains gites épicrétacés. <*Bull. de la Soc. Geol. de France*, ser 2, vol. x, pp. 511-519. 1853.

LEYMERIE, A. Note sur le massif d'Ausseing et du Saboth Haute-Garonne. <*Bull. de la Soc. Geol. de France*, sér. 2, vol. x, pp. 519-528. 1853.

LOCARD, A. Description de la Mollasse marino et d'eau douce du Lyonnais et du Dauphiné. <*Arch. du Mus. de Lyon*, vol. ii. (Foraminifères, pp. 198, 199.) 1878.

LORY, C. Note sur les terrains du Dévolny (Hautes-Alpes). <*Bull. de la Soc. Geol. de France*, sér. 2, vol. x, pp. 20-33. 1853.

LOVISATO, D. Riassunto sui terreni e posterziari del Circondario di Catawzaro. <*Boll. d. R. Com. Geol. d'Italia*, ser. 2, vol. vi, pp. 87-120 1885.

LYELL, C. Quelques considerations sur la communication précédente. <*Bull. de la Soc. Geol. de France*, sér. 2, tome ix, pp. 351-354. 1852.

MANZONI, A. Tortoniano e i suoi fossili nella Provincia di Bologna <*Bollet. del. Com. Geol. d'Italia*, vol. xi, pp. 510-520. 1880.

MENKE, C. T. Synopsis Methodica Molluscorum generum omnium et specierum quæ in Museo Menkeano adservantur. Ed. 2 auct. et emend. Pyrmonti, 1830.

MASSOLONGO. Schizzo geogn. sulla valle del Progno. Verono, 1850.

MICHELOTTI, G. Saggio storico intorno dei Rizopodi caratteristici dei Terreni sopracretacei. <*Mem. Soc. Ital. d. Sci.*, xxii, p. 302, pls. i-iii. 1841.

MICHELOTTI, G. Description of the Fossils of the Miocene Strata of Northern Italy. <*Naturkundige Verhandlingen van de Hollandsche Maatschappij der wetenschappen le Haarlem*, II. Verzam., 3 Deel. Haarlem, 1847.

MURCHISON, R. I. On the Geological Structure of the Alps, Apennines and Carpathians, more especially to prove a transition from Secondary to Tertiary rocks, and the development of Eocene deposits in Southern Europe. <*Quart. Journ. Geol. Soc. Lond.*, vol. v, pp. 157-312, 1 plate. 1849.

MILNE-EWARDS, A. Compte rendu sommaire d'une exploration zoologique, faite dans la Méditerranée à bord du navire de l'etat "le Travailleur." <*Comptes Rendus*, 28 nov. and 5 Dec. 1881, pp. 876-882 and 931-936. 1881.

MONFORT, DENYS. De quelques argonautes qui restent tonjours petits, des corps pétrifiés qu'on peut rapporter en général au genre des argonautes, et des argonautes microscopiques. <*Hist. Nat. des Moll,*, vol. iv, pp. 1-46, 1 pl.

MONFORT, DENYS DE. Histoire Naturelle générale et particulière des Mollusques (faisant partie du *Buffon de Sonnini*). Paris, 1802-5.

MONFORT, DENYS DE. Conchyliologie systématique et classification méthodique de Coquilles, etc. Paris, 1808-10.

MONFORT, DENYS, and DE ROISSY, F. Histoire Naturelle générale et particulière des Mollusques, animaux sans vertèbres et à sang blanc (faisant partie du *Buffon de Sonnini*) Denys de Monfort, continuée par F. de Roissy. Paris, 1802-05.

MORTILLET, G, DE. Note sur le crétacé et le nummulitique des environs de Pistojà. <*Mil. Att.* iii, p. 459.

MUNIER-CHALMAS. Observations sur les Algues calcaires appartenant au groupe des Siphonées verticillées (Dazycladées, Harv.), et confondues avec les Foraminifères. <*Comptes Rendus*, vol. lxxxv, p. 814. 1877.

MUNIER-CHALMAS. Observations sur les Algues calcaires confondues avec les Foraminifères et appartenant au groupe des Siphonées dichotomes. <*Bull. de la Soc. Geol de France*, sér. 3, vol. vii, p. 661, wood cuts. 1879.

MUNIER-CHALMAS. Études sur les Nummulites lævigata, planulata, variolaria, irregularis, et sur les Assilina granulata et spira, etc. <*Bull. de la Soc. Geol. France*, sér. 3, vol. viii, p. 300. 1880.

MUNIER-CHALMAS. Sur les Nummulites. <*Bull. de la Soc. Geol. France*, sér. 3, vol. vii, pp. 300, 301. 1881.

MUNIER-CHALMAS. Caractères der Miliolidæ. <*Bull. de la Soc. Geol. France*, sér. 3, vol. x, pp. 424, 425. 1882.

MUNIER-CHALMAS and C. SCHLUMBERGER. Nouvelles observations sur le dimorphisme des Foraminifères. <*Comptes Rendus*, vol. xcvi, pp. 862-866, wood cuts 1-4, pp. 1598-1601, wood cuts 5-8. 1883.

MUNIER-CHALMAS, M. M., et SCHLUMBERGER. Note sur les Miliolidées trématophorées. <*Bull. de la Soc. Geol. de France*, sér. 3, vol. xiii, pp. 273-323, plates xiii, xiv. 40 wood cuts. 1885.

NICOLIS, E. Oligocene e Miocene nel Sistema del Monte Baldo. Verona, 1884.

PARETO, L. Note sur le terrain nummulitique du pied des Apennins. <*Bull. de la Soc. Geol. de France*, sér. 2, vol. xii, pp. 370-395. 1855.

PHILIPPI, R. A. Enumeratio Molluscorum Siciliæ, cum viventium tum ni tellure tertiaria fossilium, quæ ni itinere suo observavit, 4 maj., vol. i, Berolina, 1836; vol. ii, Halis Saxorum. 1844.

PICTET, F. J. Traité de Paléontologie, Foraminifères, vol. iv, pp. 476-526, plate cix. 1857.

PLANCHUS, J. Ariminensis, De Conchis minus notis liber, Venetiis. 1739.
    An edition at Rome in 1760.

PLANCUS, J. Appendix ad Phytobasanum (Fabio Colonna). Florence, 1744.

POTIEZ, VALERY LOUIS VICTOR, et Michaud, André LOUIS GASP. Galérie des Mollusques, ou Catalogue méthodique, descriptif et raisonné des Mollusques et Coquilles du Muséum de Douai. Paris, 1838-45.

PRATT, S. P. Sur la Géologie des Environs de Bayonne. <*Mem. Soc. Geol. de France*, ser. 2, vol. ii, p. 185, pls. v-viii. 1846.

RAULIN, V. Faits et considérations pour servir au classement du terrain à Nummulites. <*Bull. de la Soc. Geol. de France*, sér. 2, vol. v, pp. 114-128. 1848.

RAULIN, V. Note sur la position géologique du calcaire d'eau douce à Physes de Montolieu (Aude). <*Bull. de la Soc. Geol. de France*, sér. 2, vol. v, pp. 429-429. 1848.

RAULIN, V. Rectifications à la notice sur le classement du terrain à Nummulites. <*Bull. de la Soc. Geol. de France*, sér. 2, vol. v, pp. 433-437. 1848.

RAULIN, V. Quelques mots encore sur le terrain à *Nummulites* des Pyrnées. <*Bull. de la Soc. Geol. de France*, sér. 2, vol. vi, pp. 531-538. 1849.

RAULIN, V. Réponse aux observations critiques de M. Leymerie sur une note intitulée. Quelques mots encore sur le terrain à Nummulites. <*Bull. de la Soc. Geol. de France*, sér 2, vol. vii, pp. 644-650. 1850.

RAULIN, V. Note relative aux terrains tertiaires de l'Aquitaine. <*Bull. de la Soc. Geol. de France*, sér. 2, vol. ix, pp. 406-422. 1851.

READE, Rev. J. B. On the Animals of the chalk still found in the living state in the stomachs of Oysters. *Trans. Micr. Soc. Lond.*, vol. ii, pp. 20-24. 1844

RENEVIER, E. Notices géologiques et Paléontologiques sur les Alpes Vaudoises. II Massif de L'Oldenhorn. <*Bull. Soc. Vaud. d. Sci. Nat.*, vol. viii, pp. 273-290. 1865. III Environs de Cheville, vol. ix, pp. 108-109. 1866.

RISSO, J. A. Histoire naturelle des principales productions de l'Europe Méridionale et principalement de celles des environs de Nice et des Alpes maritimes. Paris et Strasbourg. 1826-27.

ROUAULT, A. Description des fossiles du terrain éocène des environs de Pau. <*Bull. de la Soc. Geol. de France*, sér. 2, vol. v, pp. 201-210. 1848.

ROUAULT, A. Description des fossiles du terrain éocène des environs de Pau. <*Mem. Soc. Geol. de France*, sér. 2, vol. iii, pp. 457-502. 1850.

RUTIMEYER, L. Recherches géol. et paléontol, sur les terrains nummulitiques des Alpes Vernoises. <Verhand. der Schweiz; Naturforsch-Gesell, bei ihrer Versammluug zu Solothurn; Bibliothéque univers de Génève, vol. ix, pp. 177-192. 1848.

SANDBERGER, F. Foraminiferen der alpinen Trias. <Verhand. K. K. Geol. Reich. 1868, pp. 219. 1868.

SANDER-RANG, A. Manuel de l'histoire naturelle des Mollusques et de leurs Coquilles. Paris. 1829.

SAGE, F. G. Observations propres à faire connaitre dans quelle classe on doit ranger les numismales. (Journ. de Physique, vol. xl) 1805.

SAUSSURE DE, H. B. De Voyage dans les Alpes, 2d ed. Neuchatel, 1779.

SAVI, P. ET MENEGHINI, G. Considerazioni sulla géolgia della Toscana, &c. Firenze, 1851.

SCHARDT, A. Étuds géologiques sur le Pays-d'enhaut Vaudois. <Bull. de la Soc. Vaud. Sci. Nat., vol. xx, p. 1-182. 1884.- (Faune de Foraminiferes du crétacé supérieur, p. 71.)

SCHLUMBERGER, C. Structure intime des Foraminiféres. <Assoc. franc. pour l'avancem. des Sci. (Lyon), vol. ii, pp. 562, 563. 1873.

SCHLUMBEGER, C. Sur un nouveau Pentellina. <Assoc. franc. pour l'avancem. des Sci. (Rochelle), pp. 230-232, wood cuts 63, 64. 1882.

SCHLUMBERGER, C. Note sur les Foraminifères. <Feuille des Jeunes Naturalistes, p. 30, pls. i-iii. 1882.

SCHLUMBERGER, C. Note sur quelques Foraminifères nouveaux ou peu connus du Golfe de Gascogne. <Feuille des Jeuns Naturalistes, xiii, année, pp. 21-28, pls ii, iii. 1883.

SCHLUMBERGER, C. Note sur le genre Cuneolina. <Bull. de la Soc. Geol. de France, sér. 3, vol. xi, pp. 272-273. 1883.

SCHLUMBERGER, C. Reproduction des Foraminifères. <Assoc franc. pour l'avancem. des Sci. (Nantes), vol. iv, pp. 800, 101. 1885.

SCHNEIDER, A. Ueber zwei neue Thalassicoilen von Messina. <Muller's Archiv., p. 38. 1858.

SCHWAGER, C. Saggio di una Classificazione dei Foraminiferi avuto riguardo alle loro No. 1, 2. Famiglie Naturali. <Bollet. del R. Comitato Geologico, ann. 1876, No. 11, 12. 1877.

SCHWAGER, C. Nota su alcuni Foraminiferi nuovi del tufo di Stretto presso Girgenti <Boll. R Comit. Geol. D. Italia, vol. ix, pp. 519-529, 1 plate. 1878.

SCORTEGAGNA, F. O. Nota Sopra le Nummoliti. <Ann. Sci. Lomb. Veneto, vol. xii, pp. 118-120; Atti Scienz. Ital 1842, pp. 235, 236.

SCORTEGAGNA, F. O. Sur les Nummulites; lettre à M. D'Orbigny. Padua. (Revue Zoolog.) Paris. 1846.

SCORTEGAGNA, F. O. Sur les Nummulites; lettre à M. le Professeur Alcide D'Orbigny, par M. le Docteur, F. O. Scortegagna, de Lonigo, Padua. 8vo. 1846. <Biblioth. Univ. Geneve, Mar. 1851. Sc. Phys. p. 254 1851.

SEGUENZA, G. Intorno ad un Nuovo Genere di Foraminiferi Fossili del Terreno Miocenico di Messina. <Eco Peloritano, anno, v, ser. 2, fasc. 9, 1 plate. 1859.

SEGUENZA, G. Prime Ricerche intorno ai Rizopodi fossili delle Argille Pleistoeeniche dei dintorni di Catania. <Atti dell' Accad. Gioenia Sci. Nat., ser. 2, vol. xviii, p. 85, pls. i, iii. 1862.

SEGUENZA, G. Notizie succinte intorno alla Costituzione Geologica dei Terreni Terziarii del Distretto di Messina. Parte prima. Table. Messina. 1862.

SEGUENZA, G. Descrizione dei Foraminiferi Monotalamici delle Marne Mioceniche del Distretto di Messina. Parte seconda, pls. i, ii. Messina. 1862.

SEGUENZA, G. Brevissimi Cenni intorno alla Serie Terziaria della Provincia di Messina. Lettera al Sig. Ing. L. Molino Foti· <Bollet. del R. Com. Geol. d'Italia. 1873.

SEGUENZA, G. Le Formazioni Terziarie nella Provincia di Reggio (Calabria). <Atti R. Accad. dei Lincei, ser. 3, vol. vi, pp. 1-446, pls. i-xvii. 1880.

SEGUENZA, G. Studi geologici e paleontologici sul Cretaceo medio dell'Italia meridionale. <Atti R. Accad. dei Lincei, Anno cclxxix, ser. 3, vol. xii, pp. 1-150, pls. i-xxi. 1882.

SEGUENZA, G. Della Lingulinopsis carlofortensis, Bornemann, jun. <Il Naturalista Siciliano, ann. iii, No. 5, p. 135. 1884.

SILVESTRI, O. Le Nodosarie fossili nel Terreno subapennino Italiano e viventi nei Mari d'Italia, 11 plates. Catania, 1872.

SISMONDA, E. Synopsis Methodica Animalium invertebratorum Pedemontii fossilium quae in collectione comi is St. Martino della Motta pro max. parte extant. Turin, 1842.

SISMONDA. Place le terrain nummulitique de Savoie dans le craie supérieure. <Bull. de la Soc. Geol. de France, sér. 1, vol. v, pp. 626-630. 1844.

SISMONDA, A. Note sur les dépôts à Nummulites. Bull. de la Soc. Geol. de France, ser. 2, vol. x, pp. 47-52. 1853.

SISMONDA, A. Lettre à M. Elie de Beaumont sur le terrain Nummulitique. <Bull. de la Soc. Geol. de France, sér. 2, vol xii, pp 807-808. 1855.

SISMONDA, A. Sur les deux formations Nummulitiques du Piémont (extrait d'und lettre à M. Elie de Beaumont). Bull de la Soc. Geol. de France, sér. 2, vol. xii, pp. 508-510. 1855.

SIX ACH. Le challenger et les abimes de la mer, Analyse de la note d. M. M. Murray et Renard sur les dépôts des mers profondes. · Ann. Soc. Geol. du Nord., vol. xi, pp. 313-335. 1884.

SOLDANI, A. Saggio orittografico ovuero osservazioni sopra le terre nautilitiche ed ammonitiche della Toscana. Sienna, 1780.

SOLDANI, A. Testaceographiæ et Zoophyographiæ-parvæ et microscopicæ, 2 vols, fol. Sienna, 1789-98.

STACHE, G. TERQUEM, M. O. Premier Mémoire sur les Foraminiféres du Systéme oolithique etude du Fullers-Eartto de la Moselle. Metz. Lorette editeur-Libraire, sul du Petit-Paris, 1867. <Ver. K. K. Geol. Reich., 1868, pp. 42, 43. 1868.

STOHR, E. Il terreno pliocenico dei dintorni di Girgenti. <Bollet. del. Com. Geol. d'Italia, vol. vii, pp. 451-474, table. 1876.

STOHR, E. Bericht über die Tripoli-Schichten auf Sicilien. <Zeitschr. d. deutsch Geol. Gesell, vol. xxix, pp. 638-643. 1877.

STOHR, E. Sulla posizione geologica del tufo e del tripoli nella zona solfifera di Sicilia. <Bollet. del. Com. Geol. d'Italia, vol. ix, pp. 498-518. 1878.

STUDER, TH. Ueber Foraminiferen aus den Alpinen Kreide. <Berner Mittheil. Naturf. Gesell., 1867, pp. 177-179.

SUESS, E. On the Occurrence of Fusulinæ in the Alps. <Quart. Journ. Geol. Soc. Lond., vol. xxvi, p. 3. 1870.

TALLAVIGNES Sur les terrains à Nummulites du département de l'Aude et des Pyrénées. <Bull. de la Soc. Geol. de France, sér. 2, vol. iv, pp. 1127-1144, 1162. 1847.

TALLAVIGNES. Résumé d'un mémoire sur les terrains à Nummulites du département de l'Aude et des Pyrénées. <Comptes rendus de l'Acad. des Sciences Naturelles, Paris, vol. xxv, p. 716. 1847.

TALLAVIGNES. Observations sur le mémoire de M. Raulin, intitulé; "Faits et considérations pour serur au classement du terrain à Nummulites." <Bull. de la Soc Geol. de France, sér. 2, vol. v, pp. 130-135. 1848.

TALLAVIGNES. Sur l'age du terrain nummulites des Pyrénées. <Bull. de la Soc. Geol. de France, sér. 2, vol. v, pp. 412-415. 1848.

TALLAVIGNES. Resumé d'un mémoire sur les terrains à Nummulites du département de l'Aude et des Pyrénées. <Archives des Sciences Naturelles de Geneve, vol. vi, p. 334, 1847; vol. ix, p. 322, 1848.

TALLAVIGNES. Resumé d'un mémoire sur les terrains à Nummulites du département de l'Aude et des Pyrénées. <Leinhard's neues Jahrbuch fur Geognosie, p. 366. 1848.

TCHIHATCHEFF, P. DE. Mémoire sur les terrains jurassique, crétacé et nummulitique de la Bithymè, de la Galatie et de la Paphlagonie. <Bull. de la Soc. Geol. de France, sér. 2, vol. viii, pp 280-297. 1851.

TCHIHATCHEFF, P. DE. Dépôts nummulitiques et diluviens de la presqu'ile de Thrace. <Bull. de la Soc. Geol. de France, sér. 2, vol. viii, pp. 297-316. 1851.

TERQUEM, O. Recherches sur les Foraminifères du Lias. Parts 1-3, in the *Mem. de l'Acad. imp. de Metz*, 2me ser, vol. xxxix-xliv, the remainder published separately.

    I. Foraminiferes du Lias du Departement de la Moslle, vol. xxxix, p. 563, pls. i-iv. 1858.
    II. Foraminiferes de l'etage moyen et de l'etage inferieur du Lias, vol. xlii, p. 415, pls. v., vi. 1862.
    III. Foraminiferes du Lias des Departements de la Moselle, Cote d'Or. du Rhone, de la Vienne et du Calvados, vol. xliv, p. 151, pls, vii—x. 1863.
    IV. Les Polymorphines des Departements de la Moselle, de la Cote-d'Or et de l'Indre, pls. xi.—xiv. 1864.
    V. Foraminiferes du Lias des Departements de la Moselle, de la Cote-d'Or et de l'Indre, pls. xv.—xviii. 1865.
    VI. Foraminiferes du Lias des Departements de l'Indre et de la Moselle, pls. xix.—xxii. 1866.

TERQUEM, O. Mémoires sur les Foraminifères du Système Oolithique. Part 1, in the *Bulletin de la Soc. d'Histoire Nat. du Dep. de la Moselle*, 1868; the remainder published by the author. 8vo.

    I. Etude du Fuller's-earthe de la Moselle, pp. 1-138, pls. i.—viii. Metz, 1867.
    II. Zone a Ammonites Parkinsoni de la Moselle, pp. 139-194, pls. ix—xxi. Metz, 1869.
    III. Les Genres Frondicularia, Flabellina, Nodosaria, Dentalina, etc., de la Zone a Ammonites Parkinsoni de Fontoy (Moselle), pp. 195-278. pls. xxii, xxix, Metz. 1870.
    IV. Les Genres Polymorphina, Guttulina, Spiroloculina, Triloculina et Quinqueloculina de la Zone a Ammonites Parkinsoni de Fontoy (Moselle) pp. 279-338, pls. xxx—xxxvii, Paris. 1874.

TERQUEM, O.. and G. BERTHELIN. Étude microscopique des Marnes du Lias Moyen d'Essey-lès Nancy. Zone inférieure de l'assize à Ammonites margaritatus. <*Mem. Soc. Geol. France*, ser. 2, vol. x, mén. 3, pls. xi-xx. 1875.

TERQUEM, O. Essai sur le classement des Animaux qui vivent sur la Plage et dans les Environs de Dunkerque. 1re fasc., 1875, pp. 1-54, pls. i-vi; 2me fasc., 1876, pp. 55-100, pls. vii-xii; 3me fasc., 1880, pp. 101-152, pls. xiii-xvii. 1875, 80.

TERQUEM, O. Recherches sur les Foraminifères du Bajocien de la Moselle. <*Bull. Soc. Geol. de France*, sér. 3, vol. iv, p. 447, pls. xv-xvii. 1876.

TERQUEM, O. Observations sur l'Etude des Foraminifères. *Bull. de la Soc. Geol. de France*, sér. 3, vol. iv, p 506, pl. xiii. 1876.

TERQUEM, O. Note sur les genres Dactylopora, Polytrypa, etc. *Bull. de la Soc. Geol. de France*, sér. 3, vol. vi, p. 83. 1877.

TERQUEM, O. Observations sur les Classifications proposées pour les Foraminifères. <*Bull. de la Soc. Geol. de France*, vol. vi, p. 211. 1878.

TERQUEM, O. Les Foraminifères et les Entomostracés-Ostracodes du Pliocène supérieur de l'Ile de Rhodes. *Mem. Soc. Geol. de France*, sér. 3, vol i, pp. 1-8, pls. i-xiv. 1878.

TERQUEM, O. Observations sur les Foraminifères du terrain tertiaire parisien. <*Bull. de la Soc. Geol. de France*, vol. vii, pp. 249-251. 1879.

TERQUEM, O. Observations sur quelques Fossiles des Époques Primaires. <*Bull. de la Soc. Geol. de France*, sér. 3, vol. viii, p. 414, pl. i. 1880.

TERQUEM, O. Note sur la communication de M. Berthelin. <*Bull. de la Soc. Geol. de France*, sér. 3, vol. xi, pp. 39-43. 1882.
Placentula partschiana.

TERQUEM, O. Les Foraminifères de l'Eocene des Environs de Paris, 20 pls. 4o. 1882.

TERQUEM, O. Observation sur une communication de M. Munier Chalmas (sur quelques genres de foraminifères.) <*Bull. de la Soc. Geol. de France*, ser. 3, vol. xi, pp. 13, 14. 1883.

TREQUEM, O. Cinquième mémoire sur les Foraminifères du Système Oolithique, pp. 339-406, pls. xxxviii-xlv. 1883.

TREQUEM, O. Sur un nouveau genre de *Foraminiferes* du *Fuller's—earth* de la Moselle. (Genre *Epistomina*.) <*Bull. de la Soc. Geol. de France*, sér. 3, vol. xi, pp. 37-39, 1 plate. 1883.

TERQUEM, O. Note relative à son 5e mémoire sur les *Foraminiferes* du système oolothique de la zone à Am. Parkensoni de Fontoy (Moselle.) <*Bull. de la Soc. de France*, ser. 3, vol xi, pp. 448, 449. 1883.

TERQUEM, M, O., and M. E. La Rade de Smyrne. <*Bull. Soc. Zool. de France*, vol. x, pp. 547-550. 3 wood cuts. 1885.

TERRIGI, G. I Rhizopodi fossili o Foraminiferi dei terreni Terziarii di Roma Studiati nelle sabbie gialle Plioceniche. <*Bullet. d Soc. Geog. Ital.*, fasc. x-xii. 1876.

TERRIGI, G. Fauna Vaticana a Foraminiferi delle Sabbie Gialle nel Plioceno Subapennino superiore, 4 pls. <*Atti. dell'Accad. Pontif. de Nuovi Lincei.*, ann. xxxiii, pp. 127-129; pls. i iv. 1880.

TERRIGI, G. Sulla fauna microscopica del calcare zancleano di Palo. <*Atti. dell' Reale Accad. dei Lincei*, ser. 3, vol. vi. 1882.

TERRIGI, G. Il Colle Quirinale, sua flora e fauna lacustre e terrestre, fauna microscopica marina delgi strati inferiori;—contribuzioni alla geologia del Bacino di Roma. <*Atti. dell'Accad. Pont. d. Nuovi Lincei*, vol. xxv, pp. 145-252, pls. vii-ix. 1883.

TERRIGI, G. Ricerche microscopiche fatte sopra frammenti di marna inclusi nei peperini laziali. <*Boll. d. R. Com. Geol. d'Italia.*, ser. ii, vol. vi, pp. 148-156. 1885.

TOURNOUER, M. Sur quelque affleurements des marnes nummulitiques de Bos-d'Srros. <*Actes. Soc Linn.*, sér. 3, vol. v, pp. 243-251. 1864.

TOURNOUER, R. Sur les Nummulites et une nouvelle espèce d'Echinide trouvées dans le "miocène inférieur" ou "oligocène moyen" des environ de Paris. <*Bull. Soc. Geol. de France*, sér. 2, vol. xxvi, pp. 372 373. 1870.

TOURNOUER, M. Sur le terrain nummulitique des environs de Castellanne. <*Bull. de la Soc. Geol. de France*, sér. 3, vol. xxix, pp. 707-719. 1872.

TOZZETTI, G. T. Relazioni d'alcuni viaggi fatti in diverse parte della Toscana, 2nd edit., 12 vols. Florence. 1768-79.

TOZZETTI, T. Relazioni d'alcuni viaggi fatti in diverse parte della Toscana. Firenze. 1792.

VANDEN BROECK, E. Observations sur la Nummulites planulata du Panisélien. <*Bull. de la Soc. Geol. de France*, sér. 3, vol. ii, p. 559. 1874.

VANDEN BROCK, E. Liste des Foraminifères du Golfe de Gascogne. 8vo. Bordeaux, 1875.

VASSEUR, G. Sur les terrains tertiaires de la Bretagne (Genus, Archiacina, Mun —Chal) <*Comptes Rendus*, vol. lxxxvii, pp. 1045-1050. 1878.

VASSEUR, G. Recherches géologiques sur les Terrains tertiaires de la France occidentale-stratigraphie. <*Ann. des Sci. Geol.*, vol. xiii, pp. 1-431. 1880.

VERNEUIL, E. (de). Description des coquilles fossiles recuillies en Crimée. <*Mem. d. Soc Geol. d. France*, sér. 1, vol. iii, pp. 37-69, plates 5, 6. 1838-39.
    Nummulites irregularis. N. distans N. polygyratus. N. rotularius. N. placentula.

VERNEUIL, E. DE. Sur les terrain Nummullitique du nord de l'Espagne. <*Bull. de la Soc Geol de France*, sér. 2, vol. vi, pp. 522-524. 1849.

VILLA, C. G. B. Rivista Geologica sulla Brianza. Milano, 1885.

VON ALBERTI, F. Ueberblick über die Trias mit Berucksichtigung ihres vorkommens in den Alpen. Stuttgart, 1869.

VON HANTKEN, M., in Fuchs' memoir-ueber den sogenannten "Badne Teger" auf Malta. <*Sitz d. k. Akad. Wiss. Wein*, vol lxxiii, p. 67, pl. i. 1876.

WATERS, A. W. Quelques Roches Alpes vaudoise étudiées au microscope. <*Bull Soc. Vaud. Sci. Nat.*, vol. xvi, p. 593, pl. xxiv. 1879.

ZIGNO, A. DE. Nouvelles observations sur les terrains crétacés de l'Italie septentrionale. <*Bull. de la Soc. Geol. de France*, sér. 2, vol vii, pp. 25-32. 1850.

PART V.

AUSTRO-HUNGARY, BELGIUM, DENMARK, FINLAND, GERMANY, HOLLAND, NETHERLANDS, NORWAY, SWEDEN AND SWITZERLAND.

AUSTRO-HUNGARY, BELGIUM, DENMARK, FINLAND, GERMANY, HOLLAND, NETHERLANDS, NORWAY, SWEDEN AND SWITZERLAND.

ACKERMANN, H. Ueber Tiefseeforschungen. <*Sitzungb d. naturwiss Gesell, Isis. in. Dresden*, Jahrg. 1872, p. 168; Jahrg. 1874, p 177. 1872-74.

ALTH, A. Geognostisch-palæontologische Beschreibung der nächsten Umgebung von Lemberg. Foraminifera, Krei lemergel von Lemberg. <*Haidinger's Abhand*, vol. iii, pp. 171-262-271, pl ites ix, xiii. 1850.

ALTH, A. Die Versteinerungen des Nizneover Kalksteines. <*Mojsisovies und Neumayr's Beitrage Zur Palæont. von Oesterreich-Ungarn*, vol. i, p. 183. 1881.

ANDRIAN, F. F. V., and K. M. PAUL. Die geologischen Verhältmiss der Klein Karpathen und der angrenzenden Landgebiete im nordwestlichen Ungarn. <*Jahrbuch, d K. K. Geol. Reich.*, vol. xiv, pp. 325-336. 1864.

AUERBACH, (DR.) L. Ueber die Einzelligkeit der *Amæben*. <*Siebold und Kolliker's Zeitschrift*, vol. vii, p. 365. 1856.

BATSCH, A. I. G. C. Sechs Kupfertafeln mit Conchylien des Seesandes. Jena. 1791.

BESSELS, E Haeckelina gigantea. Ein Protist der Gruppe der Monothalamien. <*Jenaische Zeitschr.*, vol. ix, p. 265. 1874.

BITTNER, A. Ueber das Alter des Tüfferer Mergels und über die Verwendbarkeit von Orbitoiden zur Trennung der ersten von der zweiten Mediterranstufe. <*Verhand. d. kk. Geol. Reichst.* 1885, pp. 225-232. 1885.

BLUMENBACH, J. F. Abbildungen naturhistorischer Gegenstände. Göttingen, 1796-1810.

BOLL, E. Geognosie der deutschen Ostseeländer zwischen Ei ler und Oder. 2 plates. Neubrandenburg. 1846.

BOLL, E. Geognostische Skizze von Meklenburg als Erläuterung zu der von der deutschen geologischen Gesellshaft hexauszugebenden geognostschen Uebersichtskarte von Deutschland. <*Zeitschr. d. Deutsch Geol. Gesell*, vol. viii, p. 436. 1851.

BOLSCHE, H. Ein neues Vorkommen von Versteinerungen in der Rauchwacke des südlischen Harz-Randes. <*Neues Jahrb. fur Min., etc..* Jahrg. 1864.

BORNEMANN, J. G. Ueber die Liasformation in der Umgegend von Göttingen und ihre organischen Einschlüsse. Berlin. 1854.

Bornemann, J. G. Die mikroskopische Fauna des Septarienthones von Hermsdorf bei Berlin. <*Zeitsch. d. Deutsh Geol. Gesell*, vol. vii, pp. 307-351, pls. xii-xxi. 1855.

Bornemann, J. G. Bemerkungen über einige Foraminiferen aus den Tertiärbildungen der Umgegend von Madgeburg. <*Zeitschr. d. Deutsch. Geol. Gesell.*, vol. xii, p. 156, pl. vi. 1860.

Bornemann, (jun.) L. G. Ueber die Foraminiferengattung Involuntina. <*Zeitschr. d. Deutsch. Geol. Gesell*, vol. xxvi, pp. 702-740, 2 plates. 1874.

Boue, A. Uber die Nummuliten Ablagerungen. Sur les dépôts nummulitiques. <*Bericht. ueber die Mittheilungen von Freunden der Naturwisssnschaften in Wien.*, vol. iii, pp. 446, ——. 1847.

Boue, A. Die Nummulitenlager éocen sein. Le terrain Nummulitique regarde comme éocène. <*Oesterreichischen Blatter fur Literatur* 14 février, 1848. Berichte ueber die Mittheilungen von Freunden der Naturwissenschaften in Wein, vol. iv, pp. 135-136. 1848.

Boué, A. Ueber Nummuliten. Sur les Nummulites. <*Oesterreichische Blatter fur Literatur*. 1848. Berichte ueber die Mittherlunge von Freunder der Naturwissenschaften in Wien, vol. iv, pp. 51 et 201. 1848.

Brady, H. B. Une Vraie Nummulite carbonifère. Traduit par Ernest Vanden Broeck. Traductions et Reproductions pub. par la Soc. Malac. de Belgique. 1874.

Breyn, J. P. Dissartatio physica de Polythalmüs, nova Testaceorum classe. 1732.

Bronn, H. G. System der urweltlichen Konchylien durch Diagnose, Analyse, u. Abbildung der Geschlechter erläutert. Fol. Heidelberg, 1824.

Bronn, H. G. System der urnweltlichen Pflanzenthiere. Heidelberg, 1825.

Bronn, H. G. Lethæa Geognostica oder Abbildung und Beschrsitung der für die Gebirgsformationen bezeichnendsten Vevsteinerungen, 2 vols., and Atlas of 47 plates. Stuttgart, 1835-8.

Bronn, H. G. Index Palæontologicus; 1, Nomenclator.—Enumerator, 3 vols. Stuttgart, 1848-49.

Bronn, H. G. Lethæa Geognostica, oder Abbildung und Beschreibung der für die Gebirgs-Formationen bezeichnendsten Versteinerungen, 3 te auflage. Stuttgart, 1851-56.

Bronn, H. G. Die Klassen und Ordnungen des Thier-Reichs, wissenchaftlich dargestellt in Wort un Bild. Erster Band. *Amorphozoen*. Leipzig und Heidelberg, 1859.

BRUNNER, C. Beitraege zur Kenntniss der Schweizerischen Nummuliten und Flysch-formationen. Documents pour servir à la connaissance de la formation nummulitique et à celle du Flysch en Suisse. <*Mittheilungen der naturforschenden Gesellschaft in Bern*, pp. 9-21, 1-848. Neues Jahrbuch für Mineral, etc. p. 364. 1848.

BUNZEL, E. Die Fauna des marinen Tegels am Porzteich bei Voitelsbrunn unweit Nicolsburg. <*Jarhb. d. K. K. Geol. Reichsanst.*, vol. xix, pp. 202-206. 1869.

BUNZEL, E. Die Foraminiferen des Tegels von Brünn. <*Verhandl. d. k. k. Geolog. Reichsanstalt*, No. 6, p. 96. 1870.

BURTIN, F. Xavier. Oryctographie de Bruxelles. Bruxelles, 1784.

BUTSCHLI, O. Bronn's Klassen und Ordnungen des Thier-Reichs, Wissenscaftlich dargestellt in Wort und Bild. 1. Bd. Protozoa. Neu bearbeitet. Leipzig und Heidelberg, 1880-82.

CAREZ, L. Observations sur la classification des *couches tertiaires* des environs de *Cassel* (Nord) <*Bull. de la Soc. Geol. de France.*, ser. 3, vol. xi, pp. 162-164. 1883.

CAREZ, L. Note sur *l'Urgonien* et le Néocomien de la vallée du Rhône. <*Bull. de la Soc. Geol. de France*, ser. 3, vol. xi, pp. 351-367. 1883.

COHN, F. Beiträge zur Entwickelungsgeschichte der *Infusorien.* <*Siebold und Kolliker's Zeitschrift*, vol. iii, p. 257; vol. iv, p. 253. 1851, 1853.

COLLIN, J. Om Limifjordens tidligere og nuværende Marine Fauna, med særligt hensyn til Bloddyrfaunaen. (*Foraminifererne*, p. 24,) 8o Kjobenhavn, 1884.

CORNET, F. L. et A. BRIART. Compte-Rendu de l Excursion faite aux environs de Ciply. <*Mem. de la Soc. Mal., de Belg.*, vol. viii, pp 21-35. 1873.

CROSSKEY, H. W., and D. ROBERTSON. Notes on the Post-tertiary Geology of Norway. <*Proc. Phil. Soc. Glasgow*, vol. vi, pp. 346-362. 1868.

CZJZEK, J. Beitrag zur Kentniss der fossilen Foraminiferen des Wiener Beckens. <*Haidingers naturwissenschaftliche Abhandlungen*, vol. ii, p. 137, plates xii, xiii. 1848.

DADAY, E. V. Ueber eine Polythalamie der Kochsalztümpel bei Déva in Siebenbürgen. <*Zeits. Wiss.*, vol. xl, pp. 465-480, pl. xxiv, and *Math. Nat. Ber. Ung.* i, p. 357. <Cf. *Ann. N. H.*, (5) xiii, p. 307, and transl. op. cit. xiv, pp. 349-362.

> Daday describes the only known non-marine Polythalamian Foraminifer (*Entzia tetrastomella.*) It has resemblances to many different families of *Foraminifera*, and "unites the imperforate with the perforate *Polythalamia*." It occurred in a salt pool near Deva, in Transsylvania The test is chitinous, with adherent small plates of quartz; sixteen chambers in a flat dextral spiral. Not seen.

DEECKE, W. Die Foraminiferen fauna der Zone des *Stephanoceras humphriesianum* im Unter-Elsass. <*Abh. Geol. Spec.*, v. Elsass-Lathringen. iv, 68 pp., 2 pls.
    Not seen.

DE LA HARPE, P. Note sur les Nummulites de la Crimée. <*Bull. Soc. Vaudoise des Sci., Nat.*, sér. 2, vol. xiii, pp. 267-272. 1874.

DE LA HARPE, P. Note sur les Nummulites Partschi et Oosteri de la II., du Calcaire du Michelsberg, pres Stockerau (Autriche), et du Gurnigelsandstein de Suisse. <*Bull. Soc. Vaud. Sci., Nat.*, vol. xvii, pp. 33-40. 1880.

DE LA HARPE, P. Note sur la distribution par couples des Nummulites éocènes. <*Bull. Soc. Vaudoise des Sci. Nat.*, vol. xvii, pp. 429-441. 1881.

DEWALQUE, F. Note sur la glauconie d'Anvers. <*Ann. de la Soc. de Belg. Mem.*, vol. ii, pp. 3-6. 1875.

DIESING, C. M. Systematische Uebersicht der Foraminiferen monostegia und Bryozoa anopisthia. <*Sitz. K. Akad. Wiss. Wien.*, 5 Heft, p. 494. 1848.

D'ORBIGNY, A. Foraminifères Fossiles du Bassin Tertiaire de Vienne (Auriche), découverts par son Excellence le Chevalier Joseph de Hauer. Paris, 1846.

DUNIKOWSKI, E. v. Ueber einige neue Nummulitenfunde in dem ostgalizischen Karpathen. <*Verh. Geol. Reichsanst*, vol. xvii, pp. 128-130. 1884.

DUNIKOWSKI, (DR.) E. v. Einige Bemerkungen über die Gliederung des westgalizischen Karpathensandsteines. <*Verhand. d. K. K. Geol. Reichsan*, 1885, pp. 238-240.

EGGER, J. G. Die Foraminiferen der Miocän Schichten bei *Ortenburg* in Nieder-Bayern. <*Neues Jahrbuch. fur Min. Geol.*, 1857, pp. 266-311, plates 5-15. 1857.

EHRENBERG, C. G. Ueber dem blossen Auge unsichtbare Kalkthierchen und Kieselthierchen als Hauptbestandtheile de Kreidegebirge. <*Berichte d. Konigl. Preuss. Akad. Wiss. Berlin*, 1838, p. 192. 1838.

EHRENBERG, C. G. Ueber die Bildung der Kreidefelsen und des Kreidemergels durch unsichtbare Organismen. *Abhandl. d. k. Akad. d. Wiss. Berlin* (for 1838), p. 59, pls. i-iv. 1838.

EHRENBERG, C. G. Eine vorläufige Uebersicht seiner Untersuchung der Schnerken-Corallen oder Polythalmien als Thiere. <*Berichte d. Konigl. Preuss. Akad. Wiss. Berlin* (1838), p. 196. 1838.

EHRENBERG, C. G. Anwendung den bisherigen Beobachtungen auf die Systematik der Polythalamien. *Abhand. der Konigl. Akad. d. Wissenschaften zu Berlin*. 1838.

EHRENBERG, C. G. Die Infusionsthierchen als vollkommne Organismen. Leipzic (?) Berlin, 1838.

EHRENBERG, C. G. Ueber die Bildung sämmetlicher Felsen bei dem Nilufer von Cahira bis Theben u. s. w. ans den mikroskopischen Kalkthierchen der europäischen Kreide. <*Berichte d. k. Preuss. ak. Wiss.*, 1839, pp. 26, 27. 1839.

EHRENBERG, C. G. Aufschluss über das Verhältniss der Polythalamien zur Jetztwelt und weitere Kenntniss ihrer Organisation. <*Berichte d. k. Preuss. Ak. Wiss.*. 1839, pp. 27-30. 1839.

EHRENBERG, C. G. Ueber Mehrere in Berlin lebend boebachtete Polythalamien der Nordsee. <*Berichte d. k. Preuss Ak. Wiss.*, 1840, pp. 18-23. 1840.

EHRENBERG. C. G. Meeres-Infusorien die zur Erläuterung räthselbafter fossiler Formen der Kerîdebildung dienen. <*Berichte d. k. Preuss Ak. Wiss.*, 1840, pp. 157-162. 1840.

EHRENBERG, C. G. Uber noch jetzt zahlreich lebende Thierarten der Kreidebildung und den Organismus der Polythalamien. <*Abhand. d. Akad. d. Wiss. zu Berlin*, 1839, pp. 81-174, 4 plates. 1841. Partly translated in Taylor's Scientific Memoirs, vol. iii, p. 319.

EHRENBERG, C. G. Das unsichtbar wirkende organische Leben-Vorlesung. Leipzig, 1842.

EHRENBERG, C. G. Hornstein des Bergkalkes von Tula. <*Berichte d. Kongl. Preuss. Akad Wiss. Berlin*, 1843, pp. 79, 106. 1843.

EHRENBERG, C. G. Neue Beobachtungen uber den sichtlichen Einfluss der mikroscopischen Meeres-Organismen auf den Boden des Elbbetts bis vor oberhalb Hamburg. <*Berichte d. Kongl. Preuss. Akad. Wiss. Berlin*, 1843, pp. 161-167. 1843.

EHRENBERG, C. G. Fortgesetzte Beobachtungen des bedeutenden Einflusses unsichtbar kleiner Organismen auf die unteren Stromgebiete, besonderes der Elbe, Jahde, Ems und Schelde. <*Berichte de Kongl. Preuss. Akad. Wiss. Berlin*, 1843, pp. 259-272. 1843.

EHRENBERG, C. G. Vorlaüfige Nachricht über das kleinste leben im Weltmeer, am Südpol und in den Meerestiefen. <*Berichte d. k. Preuss. Akad. Wiss.*, 1844, pp. 182-207. 1844.

EHRENBERG, C. G. Passatstaub und Blutregen. <*Abhan. d. k. Akad der Wissenschaften zu Berlin*, 269-460. Berlin, 1847.

EHRENBERG, C. G. Mikrogeologie; Das Wirken des unsichtbaren Kleinen Lebens auf der Erde. 2 vols, fol., 40 plates. Leipzig, 1854.

EHRENBERG, C. G. Weitere Ermittelungen über das Leben in grossen Tiefen des Oceans. <*Berichte d. Kongl. Preuss. Akad. Wiss.*, 1854, pp. 305-328. 1854.

EHRENBERG, C. G. Systematische Charakteristik der neun mikroskopischen Organismen des tiefen altantischen Oceans. <*Berichte d. Kongl. Preuss. Akad. Wiss.*, 1854, p. 236. 1854

EHRENBERG, C. G. Ueber neue Erkenntniss immer grösserer Organisation der Polythalamien durch deren urweltliche Steinkerne. <*Berichte d. Kongl. Akad. Wiss.*, 1855, p. 272. 1855.

EHRENBERG, C. G. Uber den Grünsand und Seine Erläuterung des organischen Lebens. <*Abhand. d. Akad. d. Wiss. zu. Berlin.* 1855, pp. 85-176, 8 plates. 1856.

EHRENBERG, C. G. Uber andere massenhafte mikroskopische Lebensformen der ältesten Silurischen Grauwacken—Thone bei Petersburg. <*Monatsbreichte der Konigl. Akad. der Wissenschaften zu Berlin.* p. 324. 1858.

EHRENBERG, C. G Ueber die Tiefgrund-verhältnisse des Oceans am Eingane d. Davisstrass und bei Island <*Sitz. d. phys-math kl. Monatsb. ak. Wiss Berlin.* 1862, pp. 275 315. 1862.

EHRENBERG, C. G. Mikrogeologische Studien über das Kleinste Leben der Meeres-Tiefgründe aller Zonen und dessen geologischen Einfluss. <*Abhand. d. Akad. d. Wiss. Berlin.* 1872, pp. 131-379, 12 plates. 1873.

EHRENBERG, C. G. Die zweite deutsche Nordpolarfahrt, vol. ii, pp. 437-467, 4 pls. 4to. Berlin. 1874.
    Translated with notes by T. Rupert Jones. *Arctic Manual,* 800, 1875, p. 571. 1875.

EHRLICH, C. Verschiedene Versteinerungen aus dem Nummuliten Sandsteine zu Maltsee. sur les fossiles du gres à Nummulites de Maltsee. <*Oesterreichische Blaetter fur Literatur*; 1848. Bericht ueber die Mittheilungen von Freunden der Naturwiss., in Wien, tome iv, p. 347. 1848.

ERTBORN, O. VAN. Note sur les sondages de la province d'Anvers. <*Soc. Geol. de Belg. Mem.,* vol. i, pp. 32–44. 1874.

FAUJAS de SAINT FOND, BARTHELEMY. Histoire naturelle de la Montagne de Saint Pierre de Maestricht., 4o. Paris, 1799.

FAUVERGE, H. G. Sur quelques roches et fossiles du bassin confluent du Rhône et de l'Ardèche. <*Bull. de la Soc. de France.*, sér. 2, vol. iii, pp. 11-14. 1846.

FICHTEL, LEOPOLD, A. V. Testacea Microscopica aliaque minuta ex Generibus Argonauta et Nautilus ad Naturam picta et descripta. 24 plates, Wien. 1798.

FICHTEL, LEOP. et MOLL, J. P. Carol. Testacea microscropia aliaque minuta ex generibus Argonauta et Nautilus, ad naturam delineata et descripta. 4o. Vienna. 1803.

FOLIN. — Sur la constitution des Rhizopodes reticulaires. C. R. (Compte Rendu) xcix, pp. 1127-1130. 1884.

The skeletal evolution of Reticular Rhizopods is discussed, and the following 9 tribes are distinguished:—Naked, half-naked, slimy, pasty, globigerinaceous, spicular, arenaceous, porcellanous, and vitreous. Each one of these stages, which lead up from the just described *Bathyopsis* to the last of the vitreous, is a group of organisms clothed in a particular manner and peculiar to it. Not seen.

FORSKAL, P. Descriptions Animalium, etc. Hafneœ, 1775.

FORSKAL, P. Icones rerum naturalium, quas in itinere orientali P. Forskal observavit, etc.; edidit Niebuhr. 4to, 43 plates. Copenhagen, 1776.

FRANZENAU, A. Rakosi Foraminiferāk. *Foldtani Kozlong*, 1881, pp. 83-107, (Beitrag zur Foraminiferen—Fauna der Rákoser (Budapest) Ober—Mediterraneen Stufe). 1881.

FRANZENAU, V. u. A, Heterolepa, eine neue Gattung ans der Ordnung der Foraminiferen. <*Termeszetrajzi fuzetek*, vol. viii, p. 3, 1884. A Museo Nationali Hungarico Budapestinensè vulgate Ungarisch und deutsch.
Not seen; taken from Ver. k. k. Geol. Reich. 1884, p. 323.

FRAUSCHER, C. F. Die Eocän-Fauna von Kosavin nächst Bribir im kroatischen Küstenlande. <*Verhand. d k k Geol Reichsanstalt*, 1884, pp. 58-62.

FUCHS, TH., und F. KARRER. Geologische Studien in den Tertiärbildungen des Wiener Beckens. <*Jahrbuch d. K. K. Geol. Reich*, bd. xx, pp. 113-140. 1870.

FUCHS, T. und F. KARRER. Geologische Studien in den Tertiärbildungen des Wiener Beckens. <*Jahrbuch d. K. K. Geol. Reich.*, vol. xxi, pp. 67-122. 1871.

FUCHS, T. und F. KARRER. Geologische Studien in den Tertiärbildungen des Weiner Beckens. <*Jahrbuch d. K. K. Geol. Reich.*, vol. xxiii, pp. pp. 117-136. 1873.

FUCHS, T. und F. KARRER. Geologische Studien in den Tertiärbildungen des Wiener Beckens. <*Jahrbuch d. K. K. Geol. Reich.*, vol. xxv, pp. 1-67. 1875.

FUCHS, T. Welche Ablagerungen haben wir als Tiefseebildungen zu betrachten? <*Neues Jahrb. fur Min., &c.*, Beilage-Baud ii, pp. 487-584. 1882.

FUCHS, T. Ueber einige Fossilien aus dem Tertiär der Umgabung Rohitsch-Sauerbrunn und über das Auftreten von Orbitoiden innerhalb des Miocans. <*Verhandl d. K. K. Geol. Reich.*, pp. 378-382. 1884.

FUSS, C. Fundort fossiler Foraminiferen am rothen Berge bei Mühlbach. <*Verhandl. u. Mittheil. des Siebenb. Vereins*, Jahrg. iii, No. 8. 1852.

GALEOTTI. Sur la Constitutian Geognostique de la province de Brabant. <*Mem. Couronnes par l'Ac. R, de Bruxelles*, vol. xii. 1837.

GEINITZ, H. B. Characteristik der Schichten und Petrefacten des sächsischen böhmischen Kreidegebirges. Dresden und Leipzig, 1839-1842.

GEINITZ, H. B. Die Versteinerungen des deutschen Zechsteingebriges und Rothliegenden. Dresden, 1848.

GEINITZ, H. B. Dyas, oder die Zechsteinformation und das Rothliegende, Heft I.—Die animalischen Ueberreste der Dyas. Leipzig, 1861.

GEINITZ, F. E. Die Flötzformationen Mecklenburgs. <*Archiv. d. Vereins der Freunde der Naturgeschichte in Mecklenburg*, 37 Jahr, 1883, pp. 1-151, pls. i-iii-v, and map. Güstron, 1883.

GESNER, C. De omni rerum fossilium genere gemmis. &c., Zurich, 1565.

GIEBEL, C. G. Thalamopora cribrosa. <*Zeitschr. fur ges Naturwiss*, ser. 2, vol. vii, p. 361. 1873.

GMELIN, J. F. Systema Naturæ, Linnæi, Ed. xiii, aucta reformata. Lipsiœ, 1789.

GOES, A. Om ett oceaniskt Rhizopodum reticulatum, Lituolina scorpiura, Montf., funnet i Osterjön. <*Ofvers, K. Vet. Akad. Forhdlg. Stockholm*, vol. xxxviii, pp. 33-35. 1882.

GOES, A. On the Reticularian Rhizopoda of the Caribbean Sea. <*Kongl. Svenska Vetenskaps-Akad, Handlingar*, vol. xix, p. 150, pls. i-xii. 1882.

GOES, A. Om *Fusulina cylindrica*, Fischer, fran Spetsbergen. <*Œfv. Ak. Forh.* 1883, pp. 29-35, with fig.
    Not seen.

GRAVENHORST, J. L. C. Aus der Infursorienwelt, 4o 1832.

GRONOVIUS, L. T. Zoophylacium Gronovianum, etc. Lugduni Batavorum, 1763-84.

GRUBER, A. Der Theilungsvorgang bei Euglypha alveolata. <*Jenaische Zeitschr. f. wiss Zool.*, vol. xxxv, pp. 432-439, pl. xxiii. 1880.

GÜMBEL, C. W. Die Streiterger Schwammlager und ihre Foraminiferen-Einschlüsse. <*Jahreshefte d. Vereins fur vaterlande Naturkunde in Wurttemburg.* 1861.

GÜMBEL, C. W. Die Streitberger Schwammlager und ihre Foraminiferen-Einschlüsse. <*Wurttemb. naturw. Jahreshefte*, xviii, pp. 192-238, 2 plates. 1862. (Abridge.)

GÜMBEL, C. W. Nummuliten-führende Schichten des Kressenberges und die Lethaea geognostica von Südbayern. Ebendas. 1865. S. 129.

GÜMBEL, C. W. Comatula oder Belemnites in den Nummuliten-Schichten des Kressenbergs. *N. Jahrb. fur Min. etc.* 1866. S. 563.

GÜMBEL, C. W. Skizze der Gliederung der oberen Schichten der Kreideformation (Pläner) in Böhmen. < *Neues Jahrbuch. fur Min. etc.*, 1867, pp. 795-809, plate 6. 1867.

GÜMBEL, C. W. Foraminiferen in den Cassianer u Raibler Schichten. < *Verhandl d. K. K. Geol. Reich.*, 1868, pp. 275, 276. 1868.

GÜMBEL, C. W. Ueber Foraminiferen u Ostracoden der St. Cassianer und Raibler Schichten. *Jahrb. d. K. K. Geol. Reichsanstalt in Wien.*, vol. xix, 1869, s 175. 1869, pl. ii

GÜMBEL, C. W. Vorläufige Mittheilungen über Tiefseeschlamm. <*Neues Jahrbuch Min* , 1870, pp. 753-768. 1870.

GÜMBEL, C. W. Ueber die Foraminiferen der Gosau und Belemnitellen Schishten. <*Sitzungsber der math. phys. cl. der k. b. Akad. der Wissensch.* 1870. S. 278.

GÜMBEL. Die geognostischen Verhältnisse des Ulmer Cementmergels seine Beziehungen zu dem lithographischen Schiefer und seine Foraminiferen-fauna. <*Sitzung d. math. phys. classe d. K. b. Akad. München*, vol i, pp. 38-72, 1 plate. 1871.

GÜMBEL, C. W. Die Sogenannten Nulliporen (*Lithothamnium* und *Dactylopora*), und ihre Betheiligung an der Tusammensetzung der Kalkgesteine. Erster Theil; die Nulliporendes Pflanzenreichs (*Lithothamnium*), pp. 13-52, 2 plates. Zweiter Theil; die Nulliporen des Thierreiches (*Dactyloporideae*) nebst Nachtrag zum ersten Theile pp. 229-290, 4 plates, Abh. d. II. cl. k. Ak. d. Wiss. vol. xi. 1871.

GÜMBEL, C. W. Uber zwei jurassische Vorläufer des Foraminiferen-Geschlechtes Nummulina und Orbitulites. <*Neues Jahrbuch fur Min. etc.*, 1872, pp. 241-260, plate B, 7. 1872.

GÜMBEL, C. W. Ueber Conodictyum bursiforme Etallon einer Foraminifere aus der Gruppe der Dactyloporideen. <*Sitzung d. moth-phys. classe. d. k. b. Akad. Wiss. Munchen*, vol. iii, pp. 282-294, 1 plate. 1873.

GÜMBEL, C. W. Ueber Coccoilthen in Eocänmergel vom Kressenberg und über Oolithbildung. <*N. Jahr., fur Min etc.* 1873.

GÜMBEL, C. W. Beitrage zur Keuntiness der Organisation und systematischen Stellung von *Receptaculites*. <*Abh. c. II, cl. k. ak. d. Wiss.*, xii bd. pp. 167-215, 1 plate. 1875.

GÜMBEL, C. W. Die geognostiche Durchforschung Bayerns, pp. 61-63. Munich, 1877.

HAAN, GUIL. DE. Monographiæ Ammoniteorum et Goniatiteorum Specimen, 8 maj., Lugdum Batav. 1825.

HAECKEL, E. De Rhizopodium finibus et ordinibus, 4to. 1861.

HAECKEL, E. Die Radiolsrien (Rhizopode Radiaria) Eine Monographie, 35 pls. Folio. Berlin, 1862.

HAECKEL, E. Monographie der Moneren. <*Jenaische Zeitschr. fur. Med. u. Naturwiss.*, vol. iv, pp. 64-137, pls. ii, iii, 1868. Translated by W. F. Kirby. <*Quart. Journ. Micr. Sci.*, vol. ix, new series, p. 207, pls. ix, x, London. 1868.

HAECKEL, E. Biologische Studien—Ersts Heft.; Studien uber Moneren und andere Protisten, 6 plates, Stuttgart. 1870.

HAECKEL, E. Ueber die Physemarien, Haliphysema und Gastrophysema. <*Report Brit. Assoc.* (Glasgow Meeting), p. 153. 1876.

HAECKEL, E. Biologische Studien, Zweites Heft; Studien zur Gastraea-Theorie, 14 plates, Jena. 1877.

HAECKEL, E. Das Protistenreich, Leipzig. 1878.

HAECKEL, E. Ueber die Phæodarien, eine neue Gruppe Kieselschaliger Rhizopoden. <*Sitzungsb. d. Jenaischen Gesell. fur Med. und Naturw.*, Jahrg. 1879. Translated. A new class of Rhizopoda. *Nature*, March 11, 1880, p. 449.

HACKEL, E. Orders of the Radiolaria. <*Journ. R. Micro. Soc*, ser. ii, vol. iv, pp. 246, 247. 1884.

HAEUSLER, R. Note sur une Zone à Globigérines dans les terrains jurassiques de la Suisse. <*Proc. Verb. Soc. Malac. Belg.*, vol. x, pp. ccxli-ccxliii. 1881.

HAEUSLER, R. Notes on the *Trochamminae* of the Lower Malm of the Canton Aargau (Switzerland). <*Ann., and Mag. Nat. Hist.*, ser. 5, vol. x, pp. 49-67. 1882.

HEAUSLER, R. (DR.) Notes sur les Foraminifères de la zone à Ammonites transversarius du canton d'Argovie. <*Bull. Soc. Vaud. Sc. Nat.*, vol. xviii, 88, pp. 220-228. 1882.

HAEUSLER, R. (DR.) Liste des foraminifères de la zone à Ammonites transversarius (Etage argovien I) du canton d'Argovie. <*Bull. Soc. Vaud. Sc. Nat.*, vol. xviii, 88, pp. 229, 230. 1882.

HAEUSLER, R. Additional Notes on the Trochamminae of the Lower Malm of the Canton Aargau, including Webbina and Hormosina. <*Ann., and Mag. Nat. Hist.*, ser 5, vol. x, pp. 349-357. 1882.

HAEUSLER, R. On the Jurassic Varieties of Thurammina papillata, Brady. <*Ann., and Mag. Nat. Hist.*, ser. 5, vol. xi, pp. 262-266, pl. viii. 1883.

HAEUSLER, R. Notes on some Upper Jurassic Astrorhizidæ and Lituolidæ. <*Qnart. Journ Geol. Soc. Lond.*, vol. xxxix, pp. 25-28, pls. ii, iii. 1883.

HAEUSLER, R. Die Astrorhizden und Lituoliden der Bimammatuszone. <*Neues Jahrb. fur Min. etc.* 1883, vol. i, pp. 55-61, pls. iii, iv. 1883.

HAEUSLER, R. Ueber die neue Foraminiferengattung Thuramminopsis. <*Neues Jahrb fur Min etc.*, 1883, vol. ii, pp. 68-72, pl. iv. 1883.

HAEUSLER, R. Die Lituolidenfauna der aarganischen Impressaschichten. <*Neues Jahrbuch Min. Geol. u Palaen.* iv Beilage-Bd, pp. 1-30, plates I-III. 1885.

HAGENOW, FR. V. Monographie de Kreide-versteinerungen Neuoovpommernsund Rügens. <*Neues Jahrbuch*. 1842.

HAGENOW, FR. V. In Geinitz, Grundriss der Versteinerungskunde. 1846.

HAIDINGER, W. Beobachtungen an der Grenze des Nummulitenkalkes und der Sandsteinformation in der Nache von Triest. Remarques sur la limite des calcaires à nummulites et des grès dans les environs de Trieste.

<Oesterreichische Blaetter für Literatur. 1848. Berichte ueber die Mittheilungen von Freunden der Wisssenschaften in Wien, vol. iv, p. 158. 1848.

HAHN, O. Die Meteorite (Chondrite) und ihre Organismen, 56 p. 32 plates. 1880.

> Plate 30, fig. 3. Nummulite of Kempter. The canals can be clearly seen with a loop.

HAUER, F. VON. On the Neogene Plastic Clay (Tegel) of Olmütz, Moravia. <Quart. Journ. Geol. Soc. Lond., vol. xix, pp. 15, 16. 1863.

HARTING, P. Die Macht des Kleinen, Deutsch von Dr. A. Schwarzkopf. Leipzig, 1851.

HARTING, P. De magt van het Kleine zigtbar in de vorming der korst van onzen aardbol, of oversigt van het maaksel, de geographische en de geologische verspreiding der polypen, der foraminiferen en polythalamien, der radiolarien of polycistinen en der diatomeën, 2 druk. 800, p. 258. Amsterdam, 1866.

HEBERT, M. E. On the Nummulitic Strata of Northern Italy and the Alps, and on the Oligocene of Germany. <Quart. Journ. Geol. Soc. Lond, vol. xxii, pp 19-23. 1866.

HEBERT. Note sur la couche à dents de squales deconverte à Bruxelles par M. Rutot. <Ann. de la Soc. Geol. de Belge., vol. i, pp. lxxii–lxxv. 1874.

HERTWIG, R. Ueber Mikrogromia socialis, eine colonie bildende Monothalamie des süssen Wassers. <Archiv. fur Mikr. Anat. vol. xx; Suppl. p. i, pl. i. 1874.

HERTWIG, R. Bemerkungen zur Organisation und systemateschen Stellung der Foraminiferen. <Jenaische Zeitschr. fur Naturwiss, vol x, p. 41, pl. ii. 1876.

HERTWIG, R. Zur Histologie der Radiolarien 5 pls. 4to. Leipzig, 1876.

HERTWIG, R. Der Organismus der Radiolarien, 10 pls. Jena, 1879.

HERTWIG, R. and E. LESSER. Ueber Rhizopoden und denselben nahestehende Organismen. <Archiv. fur Mikr. Anat., vol. xx; Suppl. p. 35, pl. i. 1874.

HILBER, V. Geologische Studien ni den ostgalizischen Miocän Gebieten. <Jahrbuch K. K. Geol. Reich., vol. xxxii, p. 137. 1882.

HISINGER, W. Lethæa Suecica, seu Petrificata Sueciae iconibus et characteribus illustrata, 4to. Stockholm, 1837.

JOZSEF-TOL, S. A pharmakosiderite és azúrvölgyet uj lelöhelye Sandberghegyen. Obhegy közelécen. (Az Eoczén nemely rétegetei a Nummulitok altal Jól Vannak jellemezve, pp. 7, 8.) <Zeit. d. Ungarischen Geol. Gesellschaft., xv 1885.

KARRER, F. Ueber das Auftreten der Foraminiferen in dem marinen Tegel des Wiener Beckens. <*Sitzb. d. K. Akad. Wiss. Wien.*, vol. xliii, p. 7, pl. 1. 1861.

KARRER, F. Ueber das Auftraten der Foraminiferen in dem marinen Tegel des Wiener Beckens. <*Sitzb. d. mathem.—naturw cl.* bd. xliv, pp 427-458, 2 plates. 1862.

KARRER, F. Über das Auftreten der Foraminiferen in de brakischen Schichten (Tegel und Sand) des Wiener Beckens. <*Sitzb. d. K. Akad. d. Wiss. mathem.—naturw, cl.*, vol. xlviii, pp. 72-101, pl. 1. 1863.

KARRER, F. Uber die Lagerung der Tertiärschichten am Rande des Wiener Beckens bei Mödling. <*Jahrbuch d. K. K. Geol. Reich.*, vol. xiii, pp. 30-32. 1863.

KARRER, F. Uber das Auftreten der Foraminiferen in dem Mergeln der marinen Uferbildungen (Leythakalk) des Wiener Beckens. <*Sitzb. d. K. Akad. d. mathem.—naturw, cl.*, vol. i, pp. 692-721, 2 plates. 1864.

KARRER, F. Uber das Auftreten von Foraminiferen in den älteren Schichten des Wiener Sandsteins. <*Sitzb. d. k. Akad. Wiss*, vol. lii, p. 492, pl. i, 1865.

KARRER, F. v. Zur Foraminiferenfauna in Osterreich. — Gesammelte Beiträge. <*Sitzb. d. k. Akad d. Wiss. mathem.—naturw, cl.*, vol. lv, pp. 331-368, 3 plates. 1867.

KARRER, F. v. Die miocene Foraminiferen fauna von Kostej im Banat. <*Sitzb. d. k. Akad. d. Wiss. mathem.—naturw, cl.*, vol. lviii, pp. 121-193, 5 plates. 1868.

KARRER, F. Ueber die Tertiäbildungen in der Bucht von Berchtoldsdorf bei Wien. <*Jahrbuch d. K. K. Geol. Reich*, vol. xviii, pp. 569-584, plate 15 1868.

KARRER, F. and TH. FUCHS. Geologische Studien in den Tertiärbildungen des Wiener Beckens. <*Jahrbuch d. K. K. Geol. Reich.*, vol. xix, pp. 190-206. 1869.

KARRER, F Ein neues Vorkommen von oberer Kreideformation in Leitzersdorf bei Stockerau und deren Foraminiferen—Fauna. <*Verhandl d. K. K. Geol. Reich.*, 1870, pp. 31-33 1870.

KARRER, F. Uber ein neues Vorkommen von oberer Kreideformation in Leitzersdorf bei Stockerau und deren Foraminiferen—Fauna. *Jahrbuch d. K. K. Geol. Reich.*, vol. xx, pp. 157-184, 2 plates. 1870.

KARRER, F. Uber die Foraminiferenfauna der sarmatischen stufe in den durch die neueren Brunnenbohrungen in Döbling, Grinzing, Brunn am Walle, etc., erschlossenen Tegel-Schichten. *Verhandl d. K. K. Geol. Reich.*, 1870, p. 44. 1870.

KARRER, F. and DR. JOHANN SINZOW. Uber das Auftreten des Foraminiferen-Genus *Nubecularia* im sarmatischen Sande von Kischenew. <*Sitzb. d. K. Akad. d. Wiss. Mathem.—Naturw, cl.*, vol. lxxiv, pp. 272-284. 1 plate. 1876.

KARRER, F., Geologie der Kaiser Franz-Josefs Hochquellen—Wasserleitung. Eine studie in den Tertiär—Bildungen am Westbande des alpinen Theiles der Niederung von Wien. 20 plates. <*Abhandl. d. K. K. gol. Reichsanstalt.* 1877.

KARRER, F., Die Foraminiferen der Tertiären Thone von Luzon. In Dr. R. v. Drasche's Fragmente zu einen Geologie der Insel Luzon, p. 75, pl. v., 1878.

KARRER, F., M. O. TERQUEM. Deuxième mémoire sur les Forminifères du Systéme Oolithique (zone à *Ammonites Parkinsoni*) de la Moselle. Metz; 1869. Gesch. d. Verf. <*Vererhandl, d. K. K. geol. Reich.* 1870, pp. 81, 82, 1870.

KAESTEN, H , Verzeichniss der im Rostocker academischen Museum befindlichen Versteinerungen aus dem Stenberger Gestein, 8vo. Rostock, 1849.

KAUFMANN, F. J., Der Pilatus, geologisch untersucht und beschrieben. Fünfte Lieferung Beiträge zur geologischen Kate der Schweiz, 4to, 10 plates and map. Bern, 1867.

KEFERSTEIN, C. Naturgeschichte des Erdkörpers, vol. ii; Palaontologie und Geologie. Leipzig, 1834.

KLEIN, J, T. Tentamen methodi ostracologicæ sive disposito naturalis Cochlidum et Concharum in suas classes, genera et species, iconibus singulorum generumæri incisis illustrata. 4 maj. Lugduni-Batavorum, 1753.

KOCH, A. Ueber einige neue Versteinerungen, etc , aus dem Hilsthon in Braunschweig (*Palaeontographica* von Dunker und von Meyer, vol. i). 1851.

KOCH, A. Geologische Studien aus der Ungeburg von Eperies. <*Verhandl, d. K. K. Reich.* 1868, pp. 218, 219, 1868.

KOCH, F. K. L. Uber einige neue Versteinerungen und die Perna Mulleti, Desh., aus dem Hilsthon vom Elligser Brink und von Holtensen im Braunschweig'schen. <*Palæontographica Dunker und von Meyer's*, vol. i, pp. 178-173, pl. xxiv, 1848.

KOCH, F., und W. DUNKER. Beitrager zur Kenntniss des Nordeutschen Oolithigebildes und dessen Versteinerungen. Braunschweig, 1837.

KÖLLIKER, A. Iconnes Histiologicæ, oder Atlas der vergleichenden gewebelchre, Erste Abtheilung—der feinere Bau der Protozoen. 4to. Leipzig, 1864.

KUBLER, DR. J. und H. ZWINGLI. Die Foraminiferen des Schweizerischen Jura, nach gemeirischaftlichen Forschungen mit Heinrick Zwingle, 4to. Winterthur, 1870.

KUBLER, J. and H. ZWINGLI. Mikroskopische Bilder aus der Urwelt der Schweiz. <*Neu jahrsblatt der Burgerbibliothek in winterthur.* 1886.

LACHMANN, JOHANN, ET CLAPAREDE; See Claparède.

LAFONT, A. Pour servir a la Faune de la Giron de contenant la Liste des Animaux Marins dont la présence a arcachon a été pendant les années 1867 et 1868. <*Actes de la Societe Linn. de Bordeaux*, vol. xxvi, ser. 3, vol. vi, pp. 518-531. 1868.

LANKESTER, E. R. Dredging in the Norwegian Fjords. <*Nature*, vol. xxvi, pp. 478, 479. 1882.

LEDERMULLER, M. F. Mikroskopische Gemuths und Augen, Ergotzungen 4to Beyr, 1760-61. Edition in French, 1764-68, Amusement Microscopiqeu, etc., 3 vols. 4to. Neuremberg.

LINNÆUS, C. A. Systema naturæ, sive regna tria naturæ systematice proposite per classes, ordines, genera et species (Edit. X) Holmiæ.
    Previous editions contained the Polythalamia ("Nautili") enumerated by other writers; but in the ninth Linnæus separates them into species, in the tenth he gives them specific names, and in the twelfth he attaches to them the synonyms of other authors. 1758.

LINNÆUS, C. Systema Naturæ, per Regna Tria Naturæ secundum classes, Ordines, Genera, et Species; cum characteribus, Differentiis, Synonymis, Locis; editis xii, reformata. Holmiæ, 1766.

LOUNICKI, M. Vorläufige Notiz über die ältesten ·tertiären Süsswasser und Meeresablagerungen in Ostgalizien. <*Verhand d. k. k. Geol. Reichsanstalt*, pp. 275-278. 1884.

LORIE, (DR.) J. Contributions à la Géologie des Pays Bas. I Résultats géologiques et paléontologiques des Forages de Puits à Utrecht, Goes et Gorkum. <*Arch. du Musee Teyler.*, ser. II, vol. ii, pp. 109-240. 1885.

MAITLAND, R. T. Descripti) systematis animalium Belgii septentrionalis, etc., pt. 7, p. 3, Rhizopodes. Leyden, 1851.

MARSSON, T. Die Foraminiferen der weissen Schreibkreide der Insel Rügen. <*Mittheil a. d. naturæ Verein v. Neu Vorpommern u. Rugen*, Jahrg x, pp. 115-196, pl. i.-v. 1878.

MARTENS. Reise nach Venedig, Th. ii. Ulm, 1838.
    Not seen.

MARTIN, K. Die Versteinerungens-führenden Sedimente Timors, Nach Sammlungen von Reinwardt, Macklot, und Schneider. <*Jaarboek van het Mijnwerzen in Nederlandsch Oost-Indie*, 1882, pp. 69-136, pls. ii, iii. 1882.

MARTINI, F. H. W., and CHEMNITZ, J. H. Neues systemat. Conchylien Cabinet, geordnet u. beschrieben, 11 vols., 4o. Nürnberg, 1769-95.

MARTONFI, L. Oslénytani Tanulmanyok a Foraminiferakól. 8vo. Kolozsvart. 1880.

MARTONFI, L. A. Kolozsvar vidéki harmadkori rétegek Foraminiferai (Fossil Foraminifera from the Tertiaries of Klausenberg, Transylvania), orvos-természettudományi Értesitő. Klausenburg, 1882.

MAYER, K. Classification der Foraminiferen nach Reuss, Jones und Vanden Broeck, p. 4. (Privately issued.) Zurich, 1877-78.

MERIAN, P. On the Foraminifera of the neighbourhood of Basle. <Quart. Journ. Geol. Soc. Lond., vol. viii, p. 38. 1852.

MILLER, H. J., ET E. VANDEN BROECK. Les Foraminifères vivants et fossiles de la Belgicque. <Mem. de la Soc. Mal. de Belg , vol. vii, pp. 15-46, 2 plates. 1872.

MILLER, H. J. Observations sur la *Nummulites planulata* Var A. minor d'Arch et Haim. <Bull de la Soc. Mal. de Belg., vol. vii, pp. xx, xxi. 1873.

MOEBIUS, K. Neue Rhizopoden. <Tageblatt der 49 'Versamml d. deutsch. Naturfor. in Hamburg, p. 15. 1876.

MOEBIUS, K. Ueber die Bedeutung der Foraminiferen für die Abstammungslehre. <Tageb. 53 Versamml deutsche Naturf., pp. 81, 82. 1880.

MOLLER, V. v. Ueber einige Foraminiferen führende Gesteine Persien's. <Jahrbuch d. K. K. Geol. Reich., vol. xxx, pp. 573-586, 2 plates. 1880.

MULLER, J. Ueber die *Thalassicollen, Polycystinen*, and *Acanthometren* des Mittelmeeres. <Abhan. der K. Akad. der Wiss. zu Belin., p. 1. 1858.

MURRAY, A. Diss. Fundamenta Testaceologiæ (Linnæus). Upsaliæ, 1771.

NEUGEBOREN, J. L. Foraminiferen aus dem Tegel-Thon von Felső-Lapugy, unveit Dobra. <Beiblatt zum siebenburger Boten, Jahrg. 1846, No. 94, pp. 433, 434. 1846.

NEUGEBOREN, J. L. Die ersten Ergebnisse der Untersuchungen des Herrn Kustos Neugeboren in Hermannstadt uber die Foraminiferen des Tegels von Felső-Lapugg unweit Dobra in Siebenbürgen. <Haidinger's Berichte uber die Mittheilungen, etc., vol. ii, pp. 163, 164. 1847.

NEUGEBOREN, J. L. Der Tegelthon von Ober-Lapugy und sein Gehalt an Foraminiferen-Gehäusen. <Verhandl. und Mittheil. des Siebenb. Vereins fur Naturw. zu Hermannstadt, Jahrg. I, No. 77. 1849.

NEUGEBOREN, J. L. Mittheilung über die Ergebnisse der weitern Untersuchung des Tegelthones von Ober-Lapugy. <Haidinger's Berichte uber die Mittheilungen, etc., vol. iii, pp. 256-260. 1849.

NEUGEBOREN, J. L. Foraminiferen von Felső-Lapugy unweit Dobra im Karlsburger District; beschrieben und nach der Natur gezeichnet. <Verhand. u. Mittheil. des Siebenb. Vereins fur Naturw. zu Hermannstadt, Jahrg. i-iii. 1849-52.

    1ter Art. Glandulina, Jahrg. I., Nos. 3, 4.—1 plate.
    2ter Art. Frondicularia and Amphimorphina, Jahrg. I., No. 8.—2 plates.
    3ter Art. Marginulina, Jahrg. II , Nos. 7, 8, 9.—2 plates.
    4ter Art. Nodosaria, Jahrg. III., Nos. 3, 4.—1 plate.

NEUGEBOREN, J. L. Schreib über fossile Polythalamien Siebenbürgens. <*Meyn. Kieler Monatsschrift; Wahrscheinlich in den Verhandlungen des Siebenburgischen Vereins fur Naturwissenschaften.* 1853.

NEUGEBOREN, J. L. Der Lingulina costata als einer für Siebenbürgen neuen Foraminiferen—species. < *Verhandl. u. Mittheil. des siebenb. veriens*, Jahrg. iv, No. 2. 1853.

NEUGEBOREN, J. L. Die Foraminiferen aus der Ordnung der Stichostegier von Ober-Lapugy in Siebenbürgen. <*Denkschr. d. math-naturw. cl. d. k. Akad. Wiss.*, vol. xii, p. 65, pls. i–v. 1856.

NEUGEBOREN, J. L. Berichtigungen zu den in den Jahrgangen LII and LIII, der Verhandl. und Mittheil. über Foraminiferen von Ober-Lapugy erschienenen aufsatze. < *Verhandl. u. Mittheil, d. siebend- Vereins fur Naturw.*, Jahrg. iii. 1860.

NEUGEBOREN, J. L. Die Cristellarien und Robulinen aus der Thierklasse der Foraminiferen aus dem marinen Miocan bei Ober-Lapugy in Siebenbürgen. <*Archiv. des des Vereines fur siebenburgische Landeskunde*, new series, vol. x, p. 273, pls. i.– iii. 1872.

NIEDZWIEDZKI, J. Miocan am Sudwest-Rande des Galizische-Podoischen Plateaus. <*Verhandl. d. K. K. Geol. Reich.* 1879. pp. 263-268, 1879.

NIEDZWIEDZKI, J. Bisherigebnisseder Tiefbohrung in Kossocice bei Wieliczka. <*Verhand. d. K. K. Geol. Reich.* pp. 331, 332. 1885.

NILSSON, S. Om de mangrummiga snackor som förekomma i Kritformationen i Sverige. <*Stockholm Akad. Handl.*, vol. xlvi, pp. 329-343. 1825.

NILSSON, S. Petrificata Suecana formationis cretaceæ, descripta et inconibus illustrata, pt. i, Fol. Lund. 1827.

N. J. Foraminifères Fossiles du Bassin Tertiaire de Vienne (Autriche) déconverts par *Joseph de Hauer* et décrits par *Alcide d Orbigny*. Paris, 1846, 1 vol., 4to, with 21 plates. <*Quart. Journ. Geol. Soc. Lond.*, vol. iii, part ii, pp. 69-71. 1847.

NYST, H., et E. VANDEN BROECK. Observations sur le même sujet. *Bull. de la Soc. Mal. de Belg.*, vol. viii, pp. xxi-xxv. 1873.

OLSZEWSKI, DR. ST. Zapiski paleontologiczne, (Foraminifera of the Chalk of Lemberg.) <*Schriften d. physiogr. Comm. d. k. k. Ak. Wiss. Krakau*, vol. ix, 95 plates. Cracow, 1874.

PARKER. W. K. and JONES, T. R. Description of some Foraminifera from the coast of Norway. < *Ann. and Mag. Nat. Hist.*, ser. 2, vol. xix, p. 273, 2 plates. 1857.

PAUL, K. M. Ein Beitrag zur Kenntniss der tertiaren Randbildungen des Wiener Beckens. <*Jahrbuch d. K. K. Geol. Reich.*, vol. xiv, pp. 391-395. 1864.

PETERS, K. F. Uber Foraminiferen im Dachsteinkalk. <*Jahrbuch d. K. K. Geol. Reich.*, vol. xiii, pp. 293-298. 1863.

PICTET, F. J. Matériaux pour la Paléontologie suisse. Sér. 1—Description des fossiles du terrain Aptien de la Pert du Rhône et des environs du St. Croix, par F. J. Pictet et Renevièr, pls. i-xxiii. Geneva, 1858.

PHILLIPPI, R. A. Enumeratio Molluscorum Siciliæ, cum viventium tum in tellure tertiaria fossilum, quæ in itinere suo observavit, 4to, vol. i, 1836, Berlin; vol. ii, 1844. Halle, 1836-44.

PHILIPPI, R. A. Versteinerungen in Steinsalz von Wieliezka. <*Neues Jahrb. fur Min. &c.*, 1843, pp. 568 and 569. 1843.

PHILIPPI, R. A. Beitrage zur Kenntniss der Tertairenversteinerungen des nordwestlichen Deutschlands. Kassel, 1843.

PUSCH, GEO. G. Polens Palaontologie. Stuttgart, 1837.

REICHERT, C. B. Ueber die Bewegungs-Erscheinungen an den Scheinfussen der Polythalamien. <*Monatsb. k Preuss. Ak. Wiss. Berlin*, 1862, pp. 406-426. 1862.

REICHERT, C. B. Die Sogenannte Körnchenbewegung an den Pseudopodien der Polythalamien. <*Archiv f. Naturgesch.*, vol. xxx, pp. 191-194. 1864.

REICHERT, C. B. Bemerkungen zu M. Schultze's Journal-Artikel: Reichert und die Gromien <*Archiv f. Anat. (Reich. und du Bois)*, pp. 286, 287. 1866.

REICHERT, C. B Uber die contractile Substanz (*Sarcode, Protoplasma*) und ihre Bewegungs-Erscheinungen. <*Abhand. d. Akad. d. Wiss. zu Berlin*, 1866, pp. 151-293, 7 plates 1867.

REINSCH, P. F. Notiz über die mikroskopische Fauna der mittleren und unteren frankischen Liasschichten. <*Neues Jahrb. fur Min. etc.*, pp. 176-178. 1877.

REUSS, A. E. Geognostische Skizzen aus Böhmen, vol. ii, Die Kreidegebilde des westlichen Böhmens. Prague, 1844.

REUSS, A. E. *Polythalamia*, in Geinitz, Grundriss der Versteinerungskunde. Dresden und Leipzig. 1846.

REUSS, A. E. Die Versteinerungen der böhmischen Kreideformation, 15 plates. Stuttgard, 1845-6.

REUSS, A. E. Neue *Foraminiferen* aus den Schichten des österreichischen Tertiarbeckens Wien, aus dem 1 sten Bande der Denkschriften der Math. naturwissenschaftlichen Klasse der K. Academic Wissenschaften, 1849. vol i, pp. 365, —, pls. xlvi—li.

REUSS, A. E. Die *Foraminiferen* und *Entomostraceen* des Kreidemergels von Lemberg. <*Haidinger's Naturwiss Abhandl.*, Band iv, pp. 17, —, pls. ii-vi. 1850.

REUSS, A. E. Ueber die fossilen Foraminiferen und Entomostraceen der Septarienthone der Umgegend von Berlin. <*Zeit. d. Deutsch. Geol. Gesell.* vol. iii, pp. 49-92, 4 plates. 1851.

REUSS, A. E. Beitrag zur Paläontologie der Tertiarschichten Oberschlesiens. <*Zeitschr. der Deutsch. Geol. Gesell*, vol. iii, p. 140, pls. vii, ix. 1851.

REUSS, A. E. Die Foraminiferen aus dem Septarienthon des Forts Lepold bei Stettin. Letter to Prof. Beyrich. <*Zeitschr. d. deutsch. Geol. Gesell.*, vol. iv, pp. 16-19, woodcuts. 1852

REUSS, A. E. *Foraminiferen* des Mainzer Beckens. <*Neues Jahrb. Min etc.*, Heft 6, p. 670. 1853.

REUSS, A. E. Ueber Entomostraceen und Foraminiferen im Zechstein der Wetterau. <*Schriften der Wetterauischen Gesellschaft.* 1853.

REUSS, (DR.) Uber einige Foraminiferen, Bryozoen und Entomostraceen des Mainzer Beckens. <*Neues Jahrbuch, etc.*, 1852, pp. 670-679, plate 9. 1853.

REUSS, A. E. Beiträge zur characteristik der Kreideschichten in den Ostalpen. <*Denkschriften der Math. Nat. K. K. Acad. der Wiss. zu Wien.*, vol. vii. pls. xxv-xxviii. 1854.

REUSS, A. E. Beiträge zur Geologischen Kenntniss Mährens. <*Jahrb. d. K. K. Geol. Reichsanst.*, vol. v, pp. 659-765. 1854.

REUSS, A. E. Ein Beitrag zur genauren Kenntniss der Kreidegebilde Mecklenburgs. <*Zeit. d. d. Geol. Gesell*, vol. vol. vii, pp. 261-292. 4 plates. 1855.

REUSS, A. E. Beiträge zur charakteristik der Tertiärschichten im nördlichen und mittleren Deutschland. <*Sitz. d. Kongl. ak. Wiss, Wien.*, vol. xviii, p. 197, pls. i-xii. 1856.

REUSS, A. E. Ueber die Foraminiferen von Pietzpuhl. <*Zeit. d. d. Geol. Gesell.*, vol. x, pp 433-438. 1858.

REUSS, A. E. Ueber die Verschiedenheit der chimischen Zusammensetzung der Foraminiferenschalen. <*Sitz. der Koneg. bohmischen Gesell. der Wiss.* vol. ii, in Prag. 1859, p. 78.

REUSS, A. E. Ueber Lingulinopsis, eine neue Foraminiferengattung aus dem böhm. Pläner. *Sitz. d. Konegl bohm. Gesell. d. Wiss.*, vol. i, p. 23. 1860.

REUSS, A. E. Die mariner Tertiärschten Böhmens und ihre Versteinerungen. <*Sitzb. d. mathem.—naturw., cl.*, vol xxix, pp. 207-285, 8 plates. 1860.

Reuss, A. E. Die Foraminiferen der westphälischen Kreideformation. <*Sitzb. d. mathem.—naturw. cl.*, vol. xl, pp. 147-238, 13 plates. 1860.

Reuss, A. E. Ueber die Foraminiferen aus der Familie der Peneropliden. <*Sitz. d. Konigl. bohm., Gesell. d. Wiss.*, vol. i, p. 68. 1860.

Reuss, A. E. Ueber Atoxophragmium, eine neue Foraminiferengattung. <*Sitz. d. Konigl. bohm. Gesell. d. Wiss.*, vol. ii, p. 52. 1860.

Reuss, A. E. Uber die Frondicularideen, eine Familie. der polymeren Foraminiferen. <*Sitz. d. K. bohm. Gesell. d. Wiss.*, vol. ii, p. 72. 1860.

Reuss, A. E. Die Foraminiferen des Crag von Antwerpen. <*Sitz. d. K. Akad. d. Wiss.*, vol xlii. Auch französisch erschienen. 1860.

Reuss, A. E. Abhandlungen über fossile Krabben und Monographie über Foraminiferen u. deren Schaalen-Struktur. <*Leon & Brown Jahrb*, p. 65. 1860.

Reuss, A. E. Entwurf einer Systematischen Zusammenstellung der Foraminiferen. <*Sitz. akad, der Wiss. Wien.*, vol. xliv, p. 355, 1861; (abstract in Ann. and Mag. Nat. Hist. ser. 3, vol. viii, p. 190.

Reuss, A. E. Paläontologische Beiträge II, Die Foraminiferen des Kreidetuffs von Maastricht, pp. 304-324. III, Die Foraminiferen der Schreibkreide von Rügen, pp. 324-333. IV, Die Foraminiferen des senonischen Grünsandes von New Jersey, pp. 334-342. <*Sitzb. d. mathem.—naturw, cl*, vol. xliv, pp. 304-342, 8 plates. 1861.

Reuss, A. E. Ueber die fossile Gattung *Acicularia*. <*Sitzb. d. k. akad. d. Wiss*, vol. xliii, p. 7, pl. i. 1861.

Reuss, A. E. Neuere untersuchungen: 1, Ueber die Foripflanzung der Foraminiferen. 2, Ueber eine neue Foraminiferengattung *Haplostiche*. <*Sitzb. d Konigl. bohn. Gesell d. Wiss.*, vol. i, p. 12. 1861.

Reuss, A. E. Kurze Notiz über eine neue Foraminiferengattung *Schizophora*. <*Sitzb. d. K. Akad. d. Wiss.*, vol. ii, p. 12. 1861.

Reuss, A. E. Beiträge zur Kenntniss der tertiären Foraminiferen—Fauna. <*Sitzb. d. mathem.—naturw. cl.* xlii, bd, pp. 355-370. 2 pls. 1861.

Reuss, A. E. Entwurf einer systematischen Zusammenstellung der Foraminiferen. <*Setzb. d. mathem.—naturw. cl.*, vol xliv, pp 355-396. 1862.

Reuss, A. E Die Foraminiferen—Familie der Langenideen. <*Sitzb. d. mathem —naturw. cl.*, vol. xlvi, pp. 308-342; 8 plates. 1862.

Reuss, A. E Die Foraminiferen des norddentschen Hils und Gault. <*Sitzb. d. mathem.—naturw. cl.*, vol. xlvi, pp. 5-100; 13 plates 1862.

Reuss, A. E. Die fossilen Foramimiferen, Anthozoen und Bryozoen von Oberburg in Steiermark; 10 plates. <*Denkschr. d. Kais Akad. de Wiss.*, vol. xxiii, p. 1, pls. 1-10. 1863.

Reuss, A. E. Beiträge zur Kenntniss der tertiären Foraminiferen Fauna. <*Sitzb. d. K. Akad. d. Wiss., mathem.—naturw. cl.*, vol. xlviii, pp. 36-71; 8 plates. 1863.

Reuss, A. E. Les Foraminifères du Crag d' Anvers. <*Bullet. de l'Acad. Roy. de Belg.*, sér. 2, vol. xv, p. 137, pls. i-iii, Traduction de M. Grün. 1863.

Reuss, A. E. Zur Fauna des deutschen Oberoligocäns. <*Sitzb. k. d Akad. d. Wiss. Mathem.—Naturw. cl.*, vol. l, pp. 435-482, 5 plates. 1865.

Reuss, A. E. Uber die Foraminiferen, Anthozoen und Bryozoen des deutschen Septarienthones. <*Sitzb. d. k. Akad. d. Wiss. Mathem.—Naturw cl*, vol. lii, pp. 283-286. 1865.

Reuss, A. E. Foraminiferen und Ostrakoden der Kreide am Kanara—See bei Küstendsche. <*Sitzb. d. k. Akad. d. Wiss. Mathem.—Naturn. cl.*, vol. lii, pp. 445-470, 1 plate. 1865.

Reuss, A. E. Die fossile Fauna der Steinsalzablagerung von Wieliczka in Galizien. <*Sitzb. d. k. Akad. d. Wiss. Mathem.—Naturw. cl.*, vol. lv, pp. 17-182, 8 plates. 1866.

Reuss, A. E. Paläontolgische Beitrage Foraminiferen und Ostracoden aus den Schichten von St. Cassian (pp. 101-107, plate 1.) <*Sitzb. d. k. Akad. d. Wiss. Mathem.—Naturw. cl.*, vol. lvii, pp 79-109, 3 plates. 1868.

Reuss, A. E. Zur fossilen Fauna der Oligocänschichten von Gaas. <*Sitzb. d. k. Akad. Wien.*, vol. lix, p. 446, pls. i-vi. 1869.

Reuss, A. E. Die Foraminiferen des Septarienthones von Pietzpuhl. <*Sitzb. d. k. Akad. d Wiss. Mathem.—Naturw. cl.*, vol. lxii, pp. 455-492. 1870.

Reuss, A. E. Vorläufige Notiz über zwei neue fossile Foraminiferen gattungen. <*Sitzb. d. k. Akad. d. Wiss. Mathem.—Naturw. cl.*, bd. lxiv, pp. 277-281. 1871.
   Polyphragma and Thalamopora.

Reuss, A. D. Die Bryozoen und Foraminiferen des unteren Pläner. aus Geinitz; Das Elbthal-Gebirge in Sachsen, I Theil. Cassel. 1872.

Reuss, A. E. Die Foraminiferen, Bryozoen und Ostracoden des oberen Pläner, Geinitz; Das Elbthal Gebirge in Sachsen, II Theil. Cassel. 1874.

Reuss, A E, and A. Fritsch. Verzeichniss von 100 Gypsmodellen von Foraminiferen welche unter der Leitung der Prof. A. E. Reuss und Dr. Anton Fritsch gearbeitet wurden. Prague, 1861.

Richter, R. Aus dem thüringschen Zechstein. *Zeitschrift d deutsch. Geol. Geseil.*, vol. vii, p. 526, pl. xxvi. 1855.

Roemer, F. Ad. Die Cephalopoden des Norddeutschen tertiaren Meeressandes. <*Neues Jahrb. für Min*, etc. 1838.

Roemer, F. A. Die Versteinerungen des norddeutschen Oolithengebirges, 4to. Hanover, 1839.

ROEMER, F. A. Die Versteinerungen des Norddeutschen Kreidegebirges, pt. 2, pl. xv, 4to Hannov, 1840–41.

ROEMER, F. A. Neue Kreide-Foraminiferen. <*Neues Jahrb fur Min*, etc., Jahrg 1842, p. 272, pl. vii. B. 1842.

ROLLE, F. Ueber einige neue Vorkommen von Foraminiferen, Bryozoen und Ostrakoden in den tertiären Ablagerungen Steiermarks. <*Jahrb. d. k. k. Geol. Reichsanst*, vol. vi, pp. 351–354. 1855.

RUTIMEYER, C. Ueber das Schweitzerische Nummuliten-Terrian, mit besonderer Berücksichtigung des Gebirges zwischen dem Thunersee und der Emme, 4to. Bern, 1850.

RUTIMEYER, —. Ueber das schweizerische Nummulitenterrain, etc., Inaug., Diss Bern, 1850, p. 69 u. 82 t. 37, 43–45.
    Not seen.

RUTOT. Note sur une coupe des environs de Bruxelles. <*Ann, de la Soc. de Belg. Mem.*, vol. i, pp. 49–59. 1874.

SAINT FOND, B. F. and J. D. PASTEUR. Natuurlijke Historie van den St. Pieters Berg bij Maastricht· Numismalen en Madreporen, pp. 246–255, plate xxxiv. 1802.

SANDAHL, O. Tva nya former af Rhizopoder. <*Ofvers af K. Vet. Akad. Forh*, xiv, 1857, pp. 299-303, 2 plates. 1858.

SANDBERGER. F. Die Stellung der Raibler Schichten, Entgegnung, Foraminiferen in deuselben. <*Verhandl. d. K. K. Geol. Reich.* 1868, pp. 190–192. 1868.

SARS, G. O. Undersogelser over Handangerfjordens Fauna. I.—Crustacea, etc. <*Vidensk.-Selsk. Forhandlinger*, 1871, p. 246. 1871.

SARS, G. O. Indberetninger til Departementet for det Indre om de af ham i Aarene 1864-1878, anstillede Undersogelser angaaende Saltvandsfiskerierne. Christiania, 1879.

SARS, M. Om de i Norge forekommende fossile Dyrelevninger fra Quartærperioden, et Bidrag til vor Faunas Historie. <*Universitetsporogram for foste halvaar*, 1864. Christiania, 1865.

SARS, M. Forsatte Bemærkninger over det dyriske Livs Udbredning i Havets Dybder. <*Forhan. Vid. Selsk.* 1868, pp. 246–286. 1869. List of Protozoa, pp. 248.

SHACKO, G. Ueber Vorkomman ausgebildeter Embryonen bee einer Rhizopode, Peneroplis proteus, d'Orb. <*Sitz. Gesell. naturf. Fr. Berlin*, pp. 130-132. 1882.

SCHACKO, G. Untersuchungen an Foraminiferen. I. Globigerinen Einschluss bei Orbulina. II Embry in Peneroplis proteus. III. Perforation bei Peneroplis. <*Wiegmann's Archiv. fur Naturgeschichte*, Jahrg xlix, pp. 428-454, pls. xii, xiii. 1883.

SCHAFHAEULT, K. E. V. Ueber die Nummuliten des bayer, südöstl. Gebirges mit Abbild. <*Neues Jahrbuch fur Min.*, etc., 1846, 4 Heft., p. 406.

SCHAFHAEULT, K. Das. Em., Sud-Bayerns Lethaea geognostica. Der Kressenberg und die südlichen Hochalpen mit ihren Petrefacten. Fol., with Atlas. Leipzig, 1863.

SCHAFHAEULT, C. E. v. Die Nummuliten führenden Schichten des Kressenberges als Nachtrag zum Aufsatz gleichen Titels im zweiten Hefte des neues Jahrb. fur Min., etc., 1865, Nr. 769 bis 788.

SCHLICHT, E. v. Die Foraminiferen des Septarienthones von Pietzpuhl, 38 plates, 4o. Berlin, 1870

SCHLOTHEIM, E. F. v. Die Petrefactenkunde. Gotha, 1820.

SCHMARDA. Neue formen von Infusorien, folio. 1849.

SCHMID, E. E. Ueber die kleineren organischen Formen des Zechsteinkalks von Selters in der Wetterau. <*Neues Jahrb. fur Min.*, etc., Jahrg. 1867, p. 376, pl. vi. 1867.

SCHLUTER, C. Cœlotrochium Decheni, eine Foraminifere aus dem Mitteldevon. <*Zeitschr. deutsch. geol. Gesell.*, vol. xxxi, pp. 668-675, wood cuts. 1879.

SCHMELCK, L. On Oceanic Deposits. <*Den Norske Nordhavs-Expedition*, 1876-1878. [The Norwegian North-Atlantic Expedition, 1876-1878,— IX Chemistry—pt. ii,] pp 71, 2 maps. Christiania, London, Leipzig, Paris. 1882.

SCHNEIDER, A. Beiträge zur Naturgeschichte der Infusorien. <*Müller's Archiv.* p. 191, 1854; translated in *Ann. and Mag. Nat. Hist.*, ser. 2, vol. xiv, p. 321. 1854.

SCHNEIDER, A. Beiträge zur Kenntniss der Protozoen. <*Zeitschr. f. wissench. Zool.*, vol xxx, Suppl., pp. 446-456, pl. xxi. 1878.

SCHREIBERS, K. v. Versuch einer vollständ. Conchylienkenntness nach Linné's System. Wien, 1793.

SCHROETER, J. S. Volstandige Einleitung in die Kenntniss und Geschichte der Stein und Versteinerungen, 4 vols. 1774-84.

SCHROTER, J. S. Einleitung in die Konchylien-kenntniss nach Linne. Halle, 1783-86.

SCHROTER, J. S. Ueber Kleine natürliche Ammonshörner. *Der Naturforscher*, vol. xvii. Halle, 1782.

SCHROTER, J. S. Ueber einige Entdeckingen und Beobachtungen an Schalthieren aus den linnäischen Geschlect Nautilus, aus einigen Arten von Seesande. <*Neue Litteratur und Beitrage zur Kenntniss der Naturgeschichte, sonderlich der Conchylien und der Steine*, 8vo. Leipzig, 1784.

Schultze, Max. Uber den Organismus der Polythalamien (Foraminiferen) Nebst Bemerkungen uber die Rhizopoden im Allegemeinen. S. F. 7 plates. 1854.

Schultze, M. Beobachtungen über die Fortpflanzung der Polythalamien. <*Muller's Archiv* , p. 165, 1856; abstracted in *Quart. Journ. Micro. Sci.* vol. v, p. 220, 1857.

Schultze, M. Die Gattung Cornuspira unter den Monothalamien, und Bemerkungen über die Organisation und Fortpflanzung der Polythalamien. <*Wiegmann's Archiv.*, vol. ii, p. 287, 1860; translated in *Ann. and Mag. Nat. Hist.*, vol. vii, p. 306, 1861.

Schultze, M. S. Das Protoplasma der Rhizopoden und der Pflanzenzellen. Leipzig, 1863.

Schultze, M. S. Uber Polytrema miniaceum, eine Polythalamie. <*Weigmann's Archiv. fur Naturg.*, xxix Jahrg., vol. i, p. 81, pl. viii. 1863.

Schultze, M. S. Die Körnchenbewegung an den Pseudopodien der Polythalamien. <*Archiv. f. Naturgesch.*, vol. xxix, pp. 361, 362, 1863.

Schulze, F. E. Zoologische Ergebnisse der Nordseefahrt, vom 21 Juli bis 9 September, 1872. I, Rhizopoden; II, Jahresb. d. Komm. zur Untersuch. d. deutsch. Meere in Kiel, p. 99, pl. ii. 1874.

Schulze, F. E. Rhizopodstudien. <*Archiv. fur mikros. Anat.*, vols. x–xiii. 1874-76.

> I. Ueber den Bau und die Entwicklung von Actinosphærium Eichhornii., vol. x, p. 328, pl. xxii.
> II. Raphidiophrys pallida, etc., vol. x, p. 377, pls. xxvi, xxvii.
> III. Euglypha, Quinqueloculina fusca, vol. xi, p. 394, pls. xv, vii.
> IV. Quadrula symmetrica, etc., vol. xi, p. 329, pls. xviii xix.
> V. Mastigamoeba aspera, etc., vol. xi, p. 583, pls. xxxv, xxxvi.
> VI. Ueber den Kern der Foraminifern. 2 Hyphothetischer Stammbaum der Rhizopoden, vol. xiii, p. 9, pls ii, iii.

Schwager, C., in Dittmar's Die Contorta-Zone, p. 198, pl. iii. 1864.

Schwager, C. Beitrag zur Kenntniss der mikroskopischen Fauna Jurassischer Schichten. <*Wurttemb. naturw. Jahreshefte.* vol. xxi, p. 82-151; 5 plates. 1865.

Schwager, C., in Dr. W. Waagen's—Ueber die Zone des Ammonites transversaiius, von Prof. Dr. Albert Oppel. <*Benecke's Geognostische-palaontogische Beitrage*, vol. i, Heft ii, pp. 205-318, woodcuts. 1866.

Schwager, C. Foraminiferen aus der Zone des Ammonites Sowerbyi (Unter-Oolith). <*Geognost. palaont Beit. von Bencke, Schlvenbach und Waagen*, vol. i, Heft iii, pp. 645-665, pl, xxxiv. 1867.

Schwager, C. St. C. Foraminiferen aus der Zone des *Amm. Sowerbyi* (Unter-Oolith). <*Verhandl. d K. K. Geol. Reich.* 1870, p. 248. 1870.

Schwager, C. Ueber die paläontologische Entwicklung der Rhizopoda. <*Bronn's Klassen und Ordnungen des Thir-Reichs.*, Edit. Bütschli, pp. 242-260. 1881.

SCHWEIGGER, A. F. Handbuch der Naturgeschichte der Skeletlosen ungegliederten Thiere. Leipzig, 1830.

SIEBOLD, C. T E. v. Bericht über die im Jahre 1841 und 1842, erschienenen Arbeiten in Bezug auf die Classen der Echinodermen, Acalephen, Polypen und Infusorien. < *Wiegmann's Archiv.* Jaghr. 1843, vol. II.

SPENGLER, L. Beskrivelse over nogle i Havsandet nylig opdagede Kokillier; in Nye Samling af det Kong. Danske. Viden. Selskabs Skrifter; Kiöbenhavn, vol. i. 1781.

SPENGLER, L. Schriften der naturforsch. Gesellschaft in Kopenhagen. 1793.

SPEYER, O. Die Tertiär-Fauna von Söllingen bei Jerxheim in Herzogthum *Braunschweig*, 4to. Cassel, 1864.

STACHE, G. Die Eocengebiete in Inner-Krain und Isbrien. <*Jahrbuch d. K. K. Geol. Reich.*, vol. xiv, pp. 11-114. 1864.

STACHE, G. Geologische Reisenotizen aus Istrien. < *Verhandl. d. K. K. Geol. Reichsanstalt*, 1872, p. 215. 1872.

STACHE, G. Neue Fundstellen von Fusulinenkalk zwischen Gailthal und Canalthal in Kärnthen. <*Verhandl. d. K. K. Geol. Reich.*, 1872, p. 283. 1872

STACHE, G. Neue Petrefactenkunde aus Istrien. < *Verhandl. d. K K. Geol. Reichsanstalt*, 1873, p. 147. 1873.

STACHE, G. Die Graptolithen-Schiefer am Osternig-Berge in Kärnten. <*Jahrb. d. k. k. Geol. Reich.*, vol. xxiii, p. 175. 1873.

STACHE, G. Die Paläozoischen Gebiete der Ostalpen. < *Jahrb. d. K. K. Geol. Reich.*, vol. xxiv, 1ter Absch . p. 135; 2ter Absch., p. 333. 1874.

STEINMANN, G. Uber Fossile Hydrozoen aus der Familie der Coryniden. <*Palaeontographica*, vol. xxv, p. 101, pls. xii-xv. 1878.

STEINMANN, G. Mikroscopische Thierreste aus dem deutschen Kohlenkalke Foraminiferen und Spongien. <*Zeitschr. d. deutsch Geol. Gesell.*, 1880, p. 394, pl. xix. 1880.

STEINMANN, G. Zur Kenntniss fossilie Kalkalgen (Siphoneen). ·<*Neues Jahrb. fur Min.*, &c., vol. ii, pp. 130-140, pl. v. 1880.

STEINMANN. Die Foraminiferengattung Nummoloculina. *Neues Jahrb. fur Min.*, &c., Jahrg. 1881, p. 31, pl. ii. 1881

STUR, DIONYS, VON. Bericht über die geologische Uebersichtsaufnahme des südwestlichen Siebenbürgen im sommer 1860. · *Jahrbuch, d. K. K. Geol. Reich.*, vol. xiii, [p. 33-120. 1863.
    List of Foraminifera pp. 82, 83.

STUR, D. v. Fossilien aus den neogenen Ablagerungen von Holubica bei Pieniaky, südlich von Brody im östlichen Galizien. *Jahrbuch, d. K. K. Geol. Reich*, vol. xv, pp. 278-282. 1865.

STUR, D. Beiträge zur Kenntniss der stratigraphischen Verhältnisse der marinen Stufe des Wiener Beckens. <*Jahrbuch, d. K. K. Geol. Reich.*, vol. xx, pp. 301-342. 1870.

STUR, D. Geologie der Steiermark. Gratz, 1871.

TARANEK, K. J. Bohemian Nebelidae. <*Journ. R. Micro. Soc.*, ser. ii, vol. iv, pp. 247-249, 1884, (Translation). See also *Abh. Bohm. Gesell. Wiss.*, vol. xi, (1882) 55 pp. (5 pls.)

THURMANN, J. and A. ETALLON. Lethæa Bruntrutana ou Etudes paléontologiques et stratigraphiques sur le Jura Bernois et en particulier les Environs de Porrentruy, partie 1, 4to. 1861.

TIETZE, (DR) E. v. Beiträge zur Geologie von Lykien. <*Jahrb. d. K. K. Geol. Reichs.*, vol. xxxv, pp. 283-386. 1885.

TOULA, F. Die Tiefen der See—Ein Vortrag Plate and map. Vienna, 1875.

TOULA, F. Die Tiefsee—Untersuchungen und ihre wichtigsten Resultate. <*Mittheil. d. Geogr. Gesell. in Wien.* Jahrg., 1875, No. 2, Plate and Map. 1875.

TOULA, F. Ueber Orbitoiden und Nummuliten führen—de Kalke vom Goldberg "bei Kirchberg am Wechsel." <*Jahrbuch, d. K. K. Geol. Reich.*, vol. xix, pp. 123-136. 1879.

UHLIG, V. Die Jurabildungen in der Umgebung von Brünn. <*Mojsisovics und Neumayr's Beiträge zur Palaeont. von Oesterreich-Ungarn*, vol. i, pp. 111-182, pls, xiii xvi. 1881.

UHLIG, V. Uber einige oberjurassische Foraminiferen mit agglutinirender Schale. <*Neues Jahrb. fur Min*, etc., vol. i, p. 152. 1882.

UHLIG, V. Vorkommen von Nummuliten in Ropa in West-Galizien. <*Verhandl. d. K. K. Geol. Reichsanstalt*, Jahrg. xvi, pp. 71, 72. 1883.

UHLIG, V. Uber Foraminiferen aus dem rjasan'schen Ornatenthone. <*Jahrb. d. K. K. Geol. Reichsanstalt.*, vol. xxxiii, pp 735-774, pls. vii-ix. 1883.

UHLIG, V. Uber die geologische Beschaffenheit eines Theiles der ost und mittelgalizischen Tiefebene. <*Jahrb. d. K. K. Geol. Reichsanstalt.*, vol. xxxiv, pp. 175-231, pls. ii, iii. 1884.

VANDEN BROECK, E. et H. J. MILLER. Observations sur la Nummulites planulata. <*Bull. de la Soc. Mal. de Belg.*, vol. viii, pp. 31, 32. 1873.

VANDEN BROECK, E. Quelques considérations sur la découverte, dans le calaire Carbonifère de Namur, d un Fossile Microscopique nouveau. <*Soc. Geol. de Belge Mem.*, pp. 16 27. 1874.

VANDEN BROECK, E. Note sur les sondages de la Province d'Anvers par M. O. Ertborm. <*Soc. Geol. de Belge Mem. Ann.*, vol. i, pp. 28-31. 1874.

VANDEN BROECK, E. Une vraie Nummulite carbonifère par H. B. Brady (traduit). <*Ann. de la Soc. Mal., de Belg.* 1874.

VANDEN BROECK, E. Quelques considérations sur la découverte, dans le calcaire carbonifère de Namur, d' un fossile microscopique nouveau (genre Nummulite). <*Ann. de la Soc. Geol. de Belg.* 1874.

VANDEN BROECK, E. Note sur les Foraminifères de l' Argile des Polders. <*Ann. Soc. Belg. Micros*, vol. iii. 1876.

VANDEN BROECK, E. Instructions pour la Récolte des Foraminifères vivants. <*Ann Soc. Belge de Micros.*, vol. iv, p. 5. 1878.

VANDEN BROECK, E. Notes sur les Foraminifères du littoral du Gard. Mines imp Clavel-Ballwet. <*Bullet. soc. d'Etude Sci. Nat. de Mines*, 6 Année, p. 18. 1878.

VANDEN BROECK, E. Monographie des Foraminifères carbonifères et permiens (le genre Fusulina ecepte) par H. B. Brady. <*Ann. de la Soc. de Belg*, vol. v, Bibliographie III, pp. 7-12. 1878.

VANDEN BROECK, E, and P. COGELS. Observations sur les Couches Quaternaires et Pliocènes de Merxem près d'Anvers. <*Ann. Soc. Malac. Belg.*, vol.; Bullet de Séances, p. 68. 1877.

VERBEEK, R. D. M. Geologische Notizen über die Inseln des Niederlandisch-Indischen Archipels im Allgemeinen, und über die fossilführenden Schichten Sumatra's im Besonderen, 4to. Batavia, 1880.

VINCENT, G. Materiaux pour servir a la Faune Laekenienne des environs de Bruxelles. <*Mem. de la Soc. Mal. de Belg.*, vol. viii, pp. 7-15. 1873.

VINCENT, G., et A. RUTOT. Relevé des sondages exécutés dans le Brabant par M. Van Ertborn. <*Ann. de la Soc. de Belg. Mem.*, vol. v, pp. 67-99. 1878.

VINCENT, G., et A. RUTOT. Note sur un sondage exécuté à la brasserie de la Dyle, à Malines. <*Ann. de la Soc. de Belg. Mem.*, vol. vi, pp 13-27. 1879.

VINCENT, G., et A. RUTOT. Coup d œil sur l'état actuel d'avancement des connaissances géologiques relatives aux terrains tertiaires de la Belgique. <*Ann. de la Soc. de Belg. Mem.*, vol vi, pp. 69-155. 1879.

VON DADAY, E. On a Polythalamian from the Salt-pools near Déva in Transylvania. <*Ann., and Mag. Nat. Hist.*, ser. 5, vol. xiv, pp. 349-363. 1864.

    Translation by W. S. Dallas, F. L. S., from the *Zeitschrift für Wissenschaftliche Zoologie*, vol. xl, pp. 465-480.

VON DER MARCK (Dr.) Ueber fossile Coccolithen und Orbulinen der oberen westfalischen Kreide. *Sitz. d. naturh. Ver. d. pr. Rheinl. u. Westphal*, vol. xxviii, Corr.-Bl., pp. 60-62. 1871.

VON DUNIKOWSKI, E. Nowe Foraminifery Kredowego Marglu Lwowskiego. <*Kosmos*, pl. i. Lemberg, 1879.

Von Dunikowski, E. Die Spongien, Radiolarien und Foraminiferen der unterliassischen Schichten vom Schafberg. <Denkschr. d. math. naturn. cl. d. k. Akad. d. Wiss. Wien, vol xlv, pp. 163-194, pls. i-vi. 1882.

Von Hagenow, A. E. Die Bryozoen der Maastrichter Kreide-Bildung, 4to. Cassel, 1850.

Von Hantken, M. v. Die Tertiargebilde der Gegend westlich von Ofen. <Jahrbuck d. K. K. Geol. Reich., vol. xvi, pp. 25-58. 1866.

Von Hantken, M. Akis-czelli talyag foraminiferai. <Magyar Foldt. Fursulat Munkalatai, vol. iv, p. 75, pls. i, ii 1868.

Von Hantken, M. Die geologischen Verhaltnisse des Graner Braunkohlengebietes. <Jahrb. d. k. ungar. Geol. Anstalt, vol. i, p. 1, pls. i-v. Pest, 1872.

Von Hantken, M. Der Ofner Mergel. <Jahrb. d. k. ungar Geol. Anstalt, vol. ii, p 208. 1873.

Von Hantken, M , and S. E. Von Madarasz. Katalog der auf der Wiener Weltausstellung im Jahre 1873, ausgestelten Nummuliten. Budapest, 1873.

Von Hantken, M. Neue Daten zur geologisehen und palæontoiogischen Kenntniss des südlicken Bakony. <Jahrb. d. k. ungar. geol. Anstalt., vol. iii, pp. 340-371, pls. xvi-xx. 1875.

Von Hantken, M. Die Fauna der Clavulina Szabói Schichten, 1 Theil— Foraminiferen. <Jahrb. d. k. ungar. geol. Anstalt., vol. iv, p 1, pls. i-xvi. 1875.

Von Hantken, M. Catalogue des Nummulites à Exposition de Paris. 1878.

Von Hantken, M. Die Mittheilungen der Herrn Edm. Hébert und Munier-Chalmas über die ungarischen alttertiaren Bildungen. <Literar. Bericht. aus. Ungarn, Jahrg. iii, pp. 687 719, pls. i, ii. 1879.

Von Ronoz, Z. Calcituba polymorpha, nov. gen , nor. spec. <Sitz. d. k. Ak. Wiss. Wien., vol. lxxxviii, pp. 420-432, 1 plate. 1883.

Von Schlotheim, E. F. Beitrage zur Naturgeschichte der Versteinerungen in geognostischer Hinsicht. <Leonhard's Taschenbuch, vol. vii, pp. 1-134. Frankfort, 1813.

Von Schauroth, K. F. Ubersicht der geog. Verhaltnisse der Gegend von Recoaro im Vicentinischen. <Sitz. d. k. Ak. Wiss. Wien., vol. xvii, pp. 481-562, pls. i-iii, aud Map. 1855.

Walch, J. E. I. Die Naturgeschichte der Versteinerungen zun Erlauterung der knorrischen Sammlung von Merkwurdigkeiten der Natur., 4 vols., fol. Suremberg. 1768-73.

> French translation, 1777-78, Recueil des Monuments des Catastrophes que le Globe Terrestre a essuiees, contenant des Petrifactions dessinees et enluminees d'apres les originaux, avec l'histoire naturelle de ces corps. 4 vols fol. Nuremberg.

WALCAH UND KNORR. Sammlung von Merkwürdigteiten der Natur, etc. 1771.

WATERS, A. W. Remarks on Fossils from Oberburg, Styria. <*Quart. Journ. Geol. Soc. Lond.*, vol. xxx, pp. 337-341. 1874.

WOLF, H. v. Die Stadt Oedenburg und ihre Umgebung. <*Jahrbuch d. k. k. Geol. Reich.*, vol. xx, pp. 15-61. 1870.

WINTHER, G. Fortegnelse over de i Danmark levende Foraminiferer. <*Naturhistorisk Tideskrift*, 3 R, 9 B, p. 101. 1874.

WRISBERG. Obser de Animalculis Infusorüs, 1765. *Folding plate, 14 micro. figures.*
Not seen.

ZITTEL, K. A. Die obere Nummuliteuformation in Ungarn. <*Sitzungsb. d. K. Ak. Wiss. Wien.*, vol. xlvi, p. 353, pls i-iii. 1862.

ZITTEL, (DR.) On the Upper Nummulitic Strata of Hungary. <*Quart. Journ. Geol. Soc. Lond.*, vol. xix, p. 8. 1863.

ZITTEL, K. A. Ueber Radiolarien der oberen Kreide. <*Zeitschr. d. deutschen Geolog. Gesellsch.*, 1876, 130, Bd. 28, S. 75.

ZITTEL, K. A. Ueber fossile Spongien und Radiolarien. <*Neues Jahrb. fur Min.* 1876.

ZITTEL, K. A. Handbuch der Paläontologie unter Mitwirkung von W. Ph. Schimper München. Oldenbourg, 1876. I Bd. 1 Lieferung.

ZSIGMONDY, W. Der artesische Brunnen im Stadtwäldchen zu Budapest. <*Jahrb. d. K. K. Geol. Reichsanstalt*, vol. xxviii, p. 659. 1878.

PART VI.

RUSSIA AND TURKEY.

RUSSIA AND TURKEY.

ABICH, H. Verglichende Grundzüge der Geologie der Kaukasus, wie des armenischen und nordpersischen Gebirge. <*Mem. d. l' Acad. Imp. Sci. St. Petersbourg*, ser. 6, vol. vii, p. 528. 1858.

ABICH, H. Ueber das Steinsalz und seine geologische Stellung im russischen Armenien Paläont Theil. <*Mem. Acad. Imp. Sci. St. Petersbourg*, vol. ix, p. 61, pls. i-x. 1859.

ABICH, H. Geologische Forschungen in den Kaukasischen Ländern. II Theil,—Geologie des armenischen Hochlandes I Westhälfte, 4to, Atlas, 19 plates, map, &c. Vienna, 1882.

DUCAN, P. M. Karakoram Stones or Syringosphæridæ. <*Scientific Results of the Second Yarkand Mission*, 4 plates, 4to. Calcutta, 1879.

EHRENBERG, C. G. Bergkalk am Onega See in Russland zum Theil ganz aus sehr deutlich erhaltenen Polythalamien bestehend. <*Berichte d. Kongl. Preuss, Ak. Wiss.*, 1842, pp. 273-275. 1842.

EHRENBERG, C. G. Ueber den Gehalt an unsichtl kleinen Lebensformen aus einigen von Hrn. Prof. Koch aus Constantinopel eingesandten Proben der Meeresablagerungen in Marmora Meer und in Bosporus. <*Berichte d. Kongl. Preuss. Akad. Wiss.*, Berlin 1843, pp. 253-257. 1843.

EHRENBERG, C. G. Ueber die obersilurischen und devonischen mikroskopischen Pteropoden, Polythalamien und Crinoiden bei Petersburg in Russland. <*Sitz. d. Phys.—Math. Kl. Monatsb. Ak. Wiss. Berlin*, 1862, P. 599, pl. i. 1862.

EICHWALD, E. Zoologia Specialis, etc., vol. ii, pp. 21-25. 1829-31.

EICHWALD, E. Lethæa Rossica, ou Paléontologie de la Russie, 5 vols., 8vo, and atlas 4to Stuttgart, 1855-61.

FISCHER DE WALDHEIM, G. Adversaria Zoologica, 4to, 7 plates. Moscow, 1819.

FISCHER DE WALDHEIM, G. Uber Fusulina. <*Bull. de la Soc. Imp. des Nat. de Moscow*, vol. i, p. 329. 1829.

FISCHER DE WALDHEIM, G. Oryctographie du Gouvernement de Moscou Fol. Moscow, 1829-37.

GREWINGK, C. Die geognostischen und orographischen Verhaltnisse des nördlichen Persiens. <*Verhandl. k. k. Mineralog. Gesellsch. St. Petersburg*, p 208 ; woodcuts.

GRIMM, O. A. (The Caspian Sea and its Fauna, pt 1). St. Petersburg, 1876.

KEYSERLING, C. Bemerkungen über einige Structurverhaltnisse der Nummuliten. <Verhandlungen der kais, russiscn. mineralog. Gesellschaft zu Petersburg. 1847.

MARESCHKOWSKY, K. S. Studien über die Protozoen des nördlichen Russland, Russisch. 133 p. u 3 Taf. St. Petersburg.
Not seen.

MOLLER, V. v. Die Spiral-gewundenen Foraminiferen des Russischen Kohlenkalks. <Mem. de l' Acad. des Sci. de St. Peteasburg, 7 série, vol xxv, 147 pp., 15 plates. 1878.

MOLLER, V. v. Die Foraminiferen des russ. Kohlenkalks. <Mem. Acad. des Sci. St. Petersburg, ser. 7, vol. xxvii. 1879.
Not seen.

MOELLER, V. Uber die Fusulinen und ahaliche Foraminiferan—Formen des Russ. Kohlenkalks (vorlaüfige notiz). <Neues Jarbuch. fur Min Geol. u. Pal., pp. 139-146. 1877.

MOLLER, V. Die spiralgewundenen Foraminiferen des russ Kohlenkalks, U 13 Taf. St. Petersburg, 4to 1878.
Not seen.

MURCHISON, DE VERNEUIL AND DE KEYSERLING. Geology of Russia in Europe, vol. ii, Palæontology. 1845.

ROUILLIER AND VOSINSKY. Etudes progressives sur la Géologie de Moscou. (Bull. de la Soc. Imp. des Natur. de Moscow, xol. xxii), pp. 337, pl. K. 1849

ROUSSEAU. Voyage dans la Russie Méridionale, etc., sous la direction d' Anatole de Démidoff, vol ii. 1840.

SPRATT, T. On the Geology of Varna and the Neighbouring parts of Bulgaria <Quart. Journ. Geol. Soc. Lond., vol. xiii, pp. 72-83. 1857.

VON KEYSERLING, GRAF A. Bemerkungen über einige Structurverhaltnisse der Nummuliten. <Verhandl. d. K. russisch. min. Gesellschaft, Jahrg. 1847, pp. 16-22. 1847.

VON MERESCHKOWSKY, C. Studien über Protozen des nordlichen Russland. <Archiv fur mikroskop. Anatomie, vol. xvi, pp 153-248, pls. x. xi. 1878.

ZBORZEWSKI, A. Recherches Microscopiques sur quelques Fossiles rares de Podolie et de Volhynie. <Nou Mem. Soc Imp. des Natur. de Moscow, vol. iii, pp 301-306, plate xxviii. 1834.

ZBOWZEWSKI, A. Raretés Microscopiques Podoliennes et Volhyniennes Microphytozoa Nou. Mem. Soc. Imp des Natur. de Moscow, vol. iii, pp. 307-312. 1834.

PART VII.

AFRICA AND ASIA.

AFRICA AND ASIA.

BACON, J. Notices of *Polythalamia*, in the sand of Sahara Desert. <*Proc. Bos. Soc. Nat. Hist.*, vol. ii, p. 164. 1848.

BAILY, W. H. Descriptions of Invertebrata from the Crimea. <*Quart. Journ. Geol. Soc. Lond.*, vol. xiv, pp. 133-161. 1868.

BARKER-WEBB, P., and BERTHELOT, J. Histoire Naturelle des I'les canaries., vol. ii, p. 123; Foraminifères par. M. D'Orbiguy. Paris, 1835-40.

BELLARDI, L. Liste des fossiles nummulitiques d'Egypte de la collection du museé de mineralogie de Turin. <*Bull. de la Soc. Geol. de France*, ser. 2, vol. viii, pp. 261-263. 1851.

BRADY, H. B. On some Fossil Foraminifera from the West-coast District, Sumatra. <*Geol. Mag.*, new series, dec. II, vol. ii, p. 532, pls. xiii, xiv. 1875.

BRADY, H. B. On some fossil foraminifera from the West-coast District, Sumatra; with two plates. <*Geol. Mag.*, pp. 532-539. 1875. Ook opgenomen in het Jaarb. Mijnwezen, 1878, 1 opblz. 166 vindt men daar de beschrijving der bolronde fusuline (schwagerina), met afbeelding op platt I, fig. 6 a, b en c.

BRADY, H. B. Ueber einige arktische Tiefsee-Foraminiferen gesammelt wahrend der österreichischungarischen Nordpol-Expedition in den Jahren 1872-74. <*Denkschr. d. k. ak. Wissensch. Wien* , vol. xliii, pp. 91-110, pls. i, ii. 1881.

CARTER, H. J. On *Foraminifera*, their Organisation and their Existence in a Fossilized State in Arabia, etc. <*Journ. Bomb. Br. Roy. Asiatic. Soc.*, vol. iii, pp. 158-183, plates viii, ix. 1848.

CARTER, H. J. On the Form and Structure of the Operculina (Operculina Arabica, Crtr.) <*Journ. Bomb. Br. Roy. Asiatic. Soc.*, vol. iv, p. 430, pl. xviii. 1852.
Same Reprinted in the *Ann., and Mag. Nat. Hist.*, ser. 2, vol. x, p. 161, pl. iv. 1852.

CARTER, H. J. Description of Orbitolites Malabarica (H.J.C.), illustrative of the Spiral and not Concentric Arrangement of Chambers in D'Orbigny's Order Cyclostègues. <*Bomb. Br. Roy. Asiatic Soc.*, vol. v, p. 142, pl. ii, A. 1853.
Same Reprinted in the *Ann., and Mag. Nat. Hist.*, ser. 2, vol. xi, pp. 425-427, pl. xvi. B. 1853.

CARTER, H. J. Descriptions of some of the large Forms of Fossilized Foraminifera in Scinde; with Observations on their Internal Structure. <*Journ. Bomb. Br. Roy. Asiatic Soc.*, vol. v, pt. 1, p. 124. 1855.
Same Reprinted in the *Ann., and Mag. Nat. Hist.*, ser. 2, vol. xi, pp. 161-171, pl. vii. 1853.

CARTER, H. J. On the true position of the Canaliferous Structure in the Shell of Fossil Alveolina (D'Orbigny). <*Ann. and Mag. Nat. Hist.*, ser. 2, vol. xiv, p. 99, pl. iii. B. 1854.

CARTER, H. J. Additional Notes on the Freshwater *Infusoria* in the Island of Bombay. <*Ann. and Mag. Nat. Hist.*, ser. 2, vol. xx, p. 34. 1857.

CARTER, H. J. On Contributions to the Geology of Western India, including Sind and Beloochistan. <*Journ. Bomb. Br. Roy. Asiatic Soc.*, vol. vi, p. 161. 1860.

CARTER, H. J. Further observations on the Structure of Foraminifera, and on the larger Fossilized Forms of Scinde, &c., including a new Genus and Species. <*Journ. Bomb. Br. Roy. Asiatsc Soc.*, vol. vi, p. 31. 1861.
    Same Reprinted in the *Ann., and Mag. Nat. Hist.*, ser 3, vol. viii, p. 309, pls. xv, xvi, xvii. 1831.

CARTER, H. J. Notes on the Freshwater *Infusoria* of the Island of Bombay. <*Ann. Mag. Nat. Hist.*, ser. 2, vol. xviii, pp. 115-221. 1865.

CARTER, H. J. Discription of a Siliceous Sand-Sponge found on the southeast coast of Arabia. <*Ann. and Mag. Nat. Hist.*, ser. 4, vol. iii, pp. 15-17. 1869.

DE GROOT, M. C. Notes on the Mineralogy and Geology of Borneo and the adjacent Islands. <*Quart. Journ. Geol. Soc. Lond.*, vol. xix, pp. 515-517. 1863.

DE LA HARPE, P. Monographie der in Ægypten und der libyschen Wüste vorkommenden Nummuliten. In Zittel's—Beitrage zur Geologie u. Palaontologie der libyschen Wuste u. der angrenzenden Gebiete, pp. 157-216, pls. xxx-xxxv. *Paleontographica*, vol. xxx. 1883.

D' ORBIGNY, ALCIDE Des. Faune des Isles Canaries. (Historie des Isles Canaries, par M. M. Barker—Webb et Bertholet.) Folio. Paris, 1839.

EHRENBERG, C. G. Verbreitung des jetzt wirkenden Kleinsten organischen Lebens in Asien, Australien und Afrika, und Bildung auch des Oolithkalkes der Juraformation aus kleinen polythalamischen Thieren. < *Berichte d. Kongl. Preuss. Adad. Wiss. Berlin.*, 1843, pp. 100, 133, 137. 1843.

EHRENBERG, C. G. Organische Kreidege-bilde in Europa und Afrika. <*Abhandl. d. K. Preuss. Akad. d. Wiss.*, (for 1844), pp. 57-97. 1844.

EHRENBERG, C. G. Ueber das Kleinste Leben an mehreren bisher nicht untersuchten Erdpunkten; mikroscopische Organismen in Portugal, Spanien, Süd-Africa, im indischen Ocean, Ganges, &c. *Berichte d. Kongl. Preuss. Akad. Wiss.*, 1845, pp. 304-322, and 357-377. 1845.

EHRENBERG, C. G. Beitrag zur Kenntniss der unterseeischen Agulhas-Bank an der Südspitze Afrikas als eines sich kundgebenden grunsandigen Polythalamien-Kalkfelsens. *Monatst. d. K. Preuss. Akad. Wiss. Berlin.*, (1863), pp. 379-394. 1863.

ETHERIDGE, R. (jun). A catalogue of Australien Fossils (including Tasmania and the Island of Timor), stratigraphically and Zoologically arranged, 8vo. Cambridge, 1878.

GEINITZ, H. B. und W. v. d. MARCK. Zur Geologie von Sumatra. <*Paleontographica*, vol. xxii, pp. 399-414. 1876.

GRANT. Memoire to illustrate a geological Map of Cutch. <*Trans. Geol. Soc. Lond.*, second series, vol. v, part ii. 1840.

HAMILTON, A. On the Foraminifera of the Tertiary Beds at Petane, near Napier. <*Trans. New Zeal. Instit.*, vol. xiii, pp. 393-396, pl. xvi. 1880.

HAMILTON, W. J. On a specimen of Nummulitic Rock from the neighbourhood of Varna. <*Quart. Journ. Geol. Soc. Lond.*, vol. xi, pp. 10, 11. 1855.

HITCHCOCK, E. Notes on the Geology of several parts of Western Asia, founded chiefly on Specimens and Descriptions from American Missionaries. <*Trans. Assoc. Amer. Geol. and Nat.*, 1840-42, pp. 340-421, plate xv. 1843.

HUGUENIN, J. Note on a Species of Foraminifera from the Carboniferous Formation of Sumatra. <*Abstracts Proc. Geol. Soc*. No. 321, p. 4. 1876.

JEFFREYS, J. G. The Post-Tertiary fossils procured in the late Arctic Expedition, with notes on some of the Recent and living Mollusca from the same expedition. <*Ann., and Mag. Nat. Hist.*, ser. 4, vol. xxii, pp. 229-241. 1877.

JONES, F. W., O. RYMER. On some Recent forms of Lagenæ from Deep-sea Soundings in the Java Seas. <*Trans. Linn. Soc. Lond*, vol. xxx, p. 45, pl. xix. 1872.

JONES, T. R., in Dr. G. A. Mantell's—Notice of the Remains of the Dinornis and other Birds, and of Fossils and Rock Specimens, recently collected by Mr. Walter Mantell in the Middle Island of New Zealand, with Additional Notes on the Northern Island. <*Quart. Journ. Geol. Soc. Lond.*, vol. vi, pp. 319-342, pls. xxviii, xxix. 1850.

JONES, T. R., in Heaphy's paper on New Zealand—Foraminifera from Orakei Creek. Auckland. <*Quart. Journ. Geol. Soc. Lond.*, vol. xvi. p. 251. 1860.

JONES, T. R. Notes on some Specimens of Nummulitic Rocks from Arabia and Egypt. <*Quart. Journ. Geol. Soc. Lond.*, vol. xxv, p. 38. 1869.

KARRER, F. Die Foraminiferen-Fauna des tertiaren Grünsandsteines der Orakei-Bay bei Auckland. <*Novara-Exped. Geol. Theil.*, vol. i, Palaont p. 71, pl. xvi. 1864.

LARTET, L. Essai sur la géologie de la Palestine et des contrées avoisinantes, etc. *Ann., des. Sci. Geol.*, vol. iii. 1869.

LARTET, L. Exploration Géologique de la Mer Morte, de la Palestine, et de l'Idumée, 4to. Paris, 1877.

MACDONALD, J. D. Observations on the Microscopic Examination of Foraminifera observed in deep-sea bottoms in the Feejee Islands. <*Ann., and Mag. Nat. Hist.*, vol. xx, 2d series, p. 195. 1857.

MANTELL, DR. G. A. On the Geology of New Zealand. <*Quart. Journ. Geol. Soc. Lond.*, vol. vi. 1850.

MANTELL, W. Sketch of the Geology of part of the Eastern Coast of the Middle Island of New Zealand. <*Quart Journ. Geol. Soc. Lond*, vol. vi, pp. 319-342, 2 plates. 1850.

MARTIN, K. Untersuchungen über die Organisation von Cycloclypeus, Carp., und Orbitoides, d'Orb. <*Niederlandisches Archiv fur Zool.*, vol v, p. 185, pls. xiii, xiv. From Junghuhu's Die Tertiarschichten auf Java Palaont, Theil, Lfg. 3. 1880.

MOBIUS, K. Foraminiferen von Mauritius. <*Beitrage zur Meeresfauna der Insel Mauritius uud der Seychellen, bearbietet von K. Mobius, F. Richters und E von Martens*, 4to, 22 plates. Berlin, 1880.

MARTIN UND WICHMANN. Sammlungen des geol. Reichs Museums in Leyden. <*Beitrag zur Geologie Asiens un Australiens* (Java, p. 105). 1881.

PARKER, W. K. On the Miliolitidæ (Agathistégues, D'Orbigny) of the East Indian Seas Part I, Miliola. <*Trans. Micr. Soc. Lond.; Quart. Journ. Micr. Sci.*, vol. vi, pp. 53-59. 1858.

PARKER, W. K. AND T. R. JONES. Note on the Foraminifera from the Bryozoan Limestone near Mount Gambier, South Australia. <*Quart. Journ. Geol. Soc., Lond.*, vol. xvi, p. 261. 1860.

ROEMER, F. Ueber eine Kohlenkalk—Fauna der Westküste von Sumatra. <*Palaentographica*, vol. xxvii, pp. 1-11, plate 1. 1880 81.
    Schwagerina Verbecki, Geinitz sp. Fusulina granum-avenae, Roem.

RICHTHOFEN, F. VON F. Uber das Vorkommen von Nummulitenformation auf Japan und den Philippinen. <*Zeit. d. d. Geol Gesell.*, vol xiv, pp. 357-360. 1862.

RICHTHOFEN, BRON. VON. On the existence of the Nummulitic formation in China. <*Amer. Journ. Sci.*, vol. i, ser. 3, pp. 110-113. 1871.

RUSSEGGER, M. On altered tertiary rocks near Cairo. *Quart. Journ. Geol. Soc. Lond*, vol. v, part ii, pp. 1-4. 1849.

SCHWAGER, C. Dr. Fossile Foraminiferen von Kar-Nikobar. · *Reise der Osterreichischen Fregatte Novara um die Erde. Geologischer Thiel.*, vol. ii, pp. 187-268, 4 plates. 1866.

SCHWAGER, C. Carbonische Foraminiferen aus China und Japan. · *Richtofen's-Beitrage zur Palaontologic von China*, pp. 107-159, pls. xv-xviii. (Dated 1883.) 1882.

SCHWAGER, C. Die Foraminiferan aus den Eocanablagerungen der libyschen Wuste und Ægyptens. In Zittel's-Beitrage zur Geologie u. Palaontologie der libyschen Wüste u. der angredzenden Gebiete; pp. 81-153, pl. xxiv-xxix. <*Palaeontographica*, vol. xxx. 1883.

SOWERBY, J. DE C. Appendix to Capt. Grant's-Memoir to illustrate a Geological Map of Cutch. <*Trans. Geol. Soc. Lond.*, 2nd ser., vol. v, part ii, pl. xxiv. 1840.

STACHE, G. Foraminiferen der tertiaren Mergel des Whaingaroa Hafens (Provinz Auckland). <*Novara-Exped., Geol. Theil.*, vol. i.—Palaont., p. 161, pls. xxi-xxiv. 1864.

STACHE, G. Fusulinenkalke aus Ober-Krain, Sumatra and Chios. <*Verhand. d. k. k. Geolog. Reichsanstalt*, No. 16, pp. 369 371. 1876.

STOLICZKA, F. Cretaceous Fauna of South India, vol. iv—Rhizopoda or Foraminifera. pp. 61, 62, pl. xii, fig. 3-5. *Mem. Geol. Survey of India*, 1872-3. 1873.

STOLICZKA, F. Description of a species of Sponges and one of Foraminifera from the Cretaceous deposits of South India. <*Mem. Geol. Sur. India Palaeon Indica*, vol. iv, pp. 59-62, plate 12. 1872-3.
  Orbitoides Faujasi (Defrance.)

VANDEN BROECK. E. On some Foraminifera from Pleistocene Beds in Ischia. <*Quart. Journ. Geol. Soc. Lond.*, vol. xxxiv, pp. 197, 198. 1878.

VERBEEK, R. D. M. Die Nummuliten des Borneo-Kalksteins. <*Neues Jahrbuch, Min.*, 1871, pp. 1-14, 3 plates. 1871.

VERBEEK, R. D. M. On the Geology of Central Sumatra. <*Geol. Mag.*, new series, dec. II, vol. ii, p. 477. 1875.

VERBEEK, R. D. M. Topographische en Geologische Beschrijving van een gedeelte van Sumatra's Westkust, 415 Batavia, 1883. Fusulina granum avenac, n. sp. p. 261.

VERNEUIL, E. P. de. Liste de Fossiles des Terrains tertiaires des environs d' Alger. <*Bull, Soc. Geol. de France.*, vol. xi, pp. 74 82. 1839.

VON FRITSCH, K. Einige eocane Foraminifern von Borneo. <*Palaeontographica*, 1878, Suppl. III, pt. i, pp. 139-146, pls. xviii, xix. 1878.

WOOD, J. E. T. On some Tertiary Deposits in the Colony of Victoria, Australia. <*Quart. Journ. Geol. Soc. Lond.*, vol. xxi, pp. 389-394. 1865.

ERRATA.

Page 177, line 10 from top—before word proper insert the.
Page 178, line 3 from top—for xxv read xv.
Page 178, line 4 from bottom—for Rocks read Limestones.
Page 179, line 15 from top—for xi read xl.
Page 179, line 16 from top—for This read The.
Page 180, line 20 from top—for S. W. read J. W.
Page 181, line 23 from top—for appearance read appearances.
Page 181, line 1 from bottom—for 66 read 68.
Page 182, line 8 from bottom—for 1886 read 1868.
Page 183, line 8 from bottom—for organic read inorganic.
Page 185, line 22 from top—for Notizer read Notizen.
Page 190, line 5 from bottom—for J. B. read J. W.
Page 191, line 4 from top—for 1871 read 1875.
Page 191, line 5 from top—for 1872 read 1876.
Page 191, line 6 from top—for Englypha read Euglypha.
Page 191, line 10 from top—for *Pascedlas* read *Pusceolus*.
Page 191, line 16 from top—for viii read vii.
Page 191, line 26 from top—for *Valulina* read *Valvulina*.
Page 191, line 27 from top—for *deceurrens* read *decurrens*.
Page 191, line 28 from top—for *plicatæ* read *plicata*.
Page 191, line 30 from top—for *Rotælii* read *Rotalia*.
Page 192, line 9 from top—for *Mantelii* read *Mantelli*.
Page 192, line 28 from top—after Pembina insert Mountain.
Page 193, line 18 from top—for Tadaissac read Tadoussac.
Page 193, line 12 from bottom—for Meridinale read Meridionale.
Page 194, line 10 from bottom—for Krede read Kreide.
Page 194, lines 12, 13 from top—for *polythalmia* read *polythalamia*.
Page 196, line 1 from top—for 2881 read 1881.
Page 196, line 14 from top—for *Lepidoiites* read *Lepidolites*.
Page 197, line 6 from top—for Om read On.
Page 197, line 1 from bottom—for Carribean read Caribbean.
Page 198, line 10 from top—for South read Southern.
Page 198, line 8 from bottom—for Foraminifera read Foraminiferen.
Page 200, line 2 from top—for Murry read Murray.
Page 201, line 11 from top—for Palaontologre read Palaontologie. Analysister read Analysirter.
Page 204, line 17 from top—for tublos read tubulosa.
Page 206, line 3 from bottom—for vii read iv.
Page 208, line 23 from top—for vii read viii.
Page 210, line 19 from top—for Poldera read Polytremata.
Page 211, line 2 from top—for Roy read Ray.
Page 212, line 24 from top—for Polythemata read Polytremata.
Page 214, line 1 from bottom—for v read iii.
Page 213, line 13 from bottom—for 1882 read 1883.
Page 214, line 4 from bottom—for House read Howse. Kirkly read Kirkby.
Page 216, line 1 from top—for Prestwick's read Prestwich's.
Page 217, line 10 from top—for Mendon read Meudon.
Page 217, line 8 from bottom—for Tumanowiczie read Tumanowiczll.

Page 218, line 10 rom top—for Kirkly read Kirkby.
Page 218, line 2 from bottom—insert pp 264-266, 1 plate. For xxi read xx.
Page 219, line 11 from bottom—for Southerndoun read Southerndown.
Page 219, line 15 from bottom—for S R read St.
Page 219, line 1 from bottom—for xvi read xxvi.
Page 220, line 3 from top—for Britanica read Britannica.
Page 220, line 9 from top—for xl read xi.
Page 221, line 6 from top—for Protozon read Protozoa.
Page 222, line 5 from top—for xl read xi.
Page 222, line 6 from bottom—insert xiv.
Page 224, line 6 from top—for Snyopsis read Synopsis.
Page 224, line 19 from bottom—for 297 read 292.
Page 224, line 13 from bottom—insert Park.
Page 224, line 11 from bottom—for D read Dr.

INDEX OF AUTHORS.

A

	PAGE.
Abich, H	286
Achiardi, A. de	234
Ackermann, H	256
Adams, G	204
Adams, J	204
Agassiz, A	190
Agassiz, L	190
Alcock, T	204, 231
Allman, G. J	204
Allman, P	204
Alth, A	256
Andrian, F. F. V	256
Anon	177, 190, 204, 234
Ansted, D. T	205
Aoust, V. D	234
Armstrong, J	205, 231
Arnold, J. W. S	190
Auerbach, L	256

B

Bachmann, I	234
Bacon, (Jr.) J	290
Bailey, J. W	190, 195
Bailey, L. W	191
Baily, W. H	290
Balkwill, F. P	205
Barker, A. E	177
Barker-Webb, P	290
Barnard, W. S	191
Barrois, C	234
Barthelemy	261
Batsch, A. I. G. C	256
Bauerman, H	205
Beaumont, É. de	234
Beccarius, J. B	234
Bellardi, L	234, 290
Bennie, J	205
Berthelin, G	234, 235, 251
Berthelot, J	290
Bessels, E	256
Beudant, F. S	235

	PAGE.
Bigsby, J. J.	177, 205
Billings, E.	191
Bittner, A	256
Blainville, H. M.	235
Blainville, H. D., de	235
Blake, J. F.	205, 226
Blake, W. P.	191
Blumenbach, J. F.	256
Boehm, G.	235
Boll, E	256
Bölsche, H.	256
Bonissent	235
Bornemann, J. G.	191, 256, 257
Bornemann, (Jr.) L. G.	235, 257
Bosc, L. A. G.	235
Boubée, N.	235
Boué, A.	235, 257
Bowdich, T. E.	205
Bowerbank, J. S.	205
Brady, H. B.	186, 191, 205, 206, 207, 208, 210, 216, 222, 224, 226, 231, 257, 290
Breyn, J. P.	257
Briart, A.	258
Broadhead, G. C.	191
Brocklesby, J.	208
Brodie, (Rev.) P. B.	208
Bronn, H. G.	257
Brookes, S.	208
Brown, J	208
Brown, (Capt.) T.	208
Bruguière, J. G.	235
Brunner, C.	258
Bryce, J.	208
Buckland, W	208
Bunzel, E	258
Burbank, L. S.	177
Burtin, F. X	258
Bury, (Mrs.)	200
Bury, P. S.	200
Bütschli, O	258
Buvignier, A	236

C

Caillaux, A.	236
Cailliaud, F	236
Capellini, G.	236
Carez, L.	258
Carpenter, W. B	177, 178, 179, 180, 209, 210, 211, 220
Carruthers, W	211
Carter, H. J.	178, 179, 211, 212, 290, 291

                    STATE GEOLOGIST.   299

                                                    PAGE.
Cattaneo, G.. 236
Catullo, A... 236
Chemnitz, J. H... 269
Chimmo, W... 212
Claparède, E... 236, 269
Clark, Wm... 212, 213
Cocks, W. P.. 213
Cogels, P.. 281
Cohn, F.. 258
Collin, J.. 258
Conrad, T. A.. 191, 192
Coppi, F... 236
Cornet, F. L... 258
Cornuel, M J... 236
Costa, E. da... 213
Costa, O. G... 236, 237
Couper, J. H... 192
Craven... 192
Credner, H.. 179, 192
Crisp, F... 192
Crosby, W. O... 192
Crosskey, H. W.................................... 213, 231, 258
Crouch, E. A... 213
Cunningham, K. M... 192
Cunningham, R. O... 192
Cuvièr, G. L. C. F... 237
Czjzek, J.. 258

                            D

Daday, E. V.. 258
D'Allard, d. S... 238
Dana, J. D.. 179, 192
D'Archiac, A. (L. V.)................................. 179, 237, 238
D'Archiac et Haime, J.. 238
D'Audebard, E. I... 241
Dawson, G. M.. 192, 193
Dawson, J. W................... 178, 179, 180, 181, 182, 186, 193, 213
Deane, H... 213
De Christofori, J.. 238
Deecke, W.. 259
De Favanne.. 238
De Folin... 214, 238
Defrance, J.. 238
De Grateleup, J. P. S.. 238
De Groot, M. C... 291
De la Harpe, P...................................... 238, 239, 259, 291
Delbos, J.. 239
Deluc, G. A.. 239
De Monnet.. 244

	PAGE.
Deshayes, G. P.	239
Deslongchamps, E.	239
De Stefani, C.	240
Dewalque, F.	259
Diesing, C. M.	359
Dillwyn, L. W.	240
Dixon, F.	213
Doderlein, P.	240
D'Orbigny, A.	193, 201, 240, 259, 291
Dujardin, F.	240, 241
Duncan, P. M.	201, 213, 286
Dunikowski, (Dr.) E. v.	259
Dunker, W.	268
Duthiers, H. L.	241

E

Edwards, A. M.	182
Egger, J. G.	259
Ehrenberg, C. G.	193, 194, 201, 241, 259, 260, 261, 286
Ehrlich, C.	261
Eichwald, E.	286
Elcock, C.	213
Eley, (Rev.) H.	213
Ertborn, O. v.	261
Etallon, A.	280
Etheridge, (Jr.) R.	213, 220, 292
Ewald, J.	241

F

Fabricius, O.	194
Faujas de S. F. B.	261
Fauverge, H. G.	241, 261
Ferry, H. de.	241
Ferussac, B.	241
Fichtel, L. A. V.	261
Fischer, P.	241
Fischer de Waldheim, G.	286
Fleming, J.	214
Folin, M. de.	214, 261, 262
Fontannes, F.	241
Forbes, E	241
Fornasini, C.	241
Fortis, C. A.	242
Fortis, J. B.	242
Forskal, P.	262
Franzenau, A.	262
Franzenau, V	262
Frauscher, C. F.	262
Fric, A. (Dr.)	182

STATE GEOLOGIST. 301

PAGE.
Fritsch, A..182, 275
Fuchs, T...242, 262
Fuss, C... 262

G

Gabb, W. M.. 194
Galeotti, H. G..194, 262
Gardner, J. S... 231
Gaudin, C. T.. 242
Geddes, P.. 214
Geinitz, F. E.. 263
Geinitz, H. B..194, 262, 263, 292
Gemmellaro, G. G... 242
Gervais, P.. 242
Gesner, C... 263
Giebel, C. G... 263
Ginanni, G. (Comte)... 242
Gmelin, J. F... 263
Göes, A... 263
Gosse, P. H... 214
Grant... 292
Gravenhorst, J. L. C... 263
Gray, J. E.. 214
Green, J.. 214
Gregorio, A. de... 242
Grewingk, C... 286
Grimm, O. A... 286
Gronovius, L. T... 263
Gruber, A... 263
Gualtieri, N.. 242
Guettard, J. E.. 242
Gümbel, C. W.. 182, 214, 242, 263, 264
Guppy, R. J. L..201, 214

H

Haan, G. de... 264
Haeckel, E..264, 265
Haeusler, R... 265
Hagenow, F. V... 265
Hahn, O..183, 265
Hall, J..183, 195
Haidinger, W..265, 266
Haime, J.. 238
Hamilton, A... 292
Hamilton, W. J...242, 292
Hamlin, F. M.. 195
Hardman, E. T... 214
Harper, L... 195
Harting, P.. 266

	PAGE.
Hauer, M	183
Hauer, T. V	266
Harvey, W. H	195
Hayden, F. V	195, 197
Hébert, E	243, 266
Herbert, E	243
Heilprin, A	195
Hencken, T. S.	201
Hertwig, R	266
Hilber, V	266
Hilgard, E. W	195
Hisinger, W	266
Hitchcock, C. H	183, 195
Hitchcock, E	292
Hitchcock, R	196
Hochstetter, R. F	183
Hochstetter, Prof. V	183
Hoeven, J. Van Der	214
Hoffmann, R	183
Homersham, C	217
Honeyman, D	196
Hooke, R	214
Hopkins, F. V	196
Howse, R	214
Huguenin, J	292
Hull, E	214
Hunt, T. S	183, 184, 186
Huxley, T. H	215
Hyndman, G. L	215

I

Issel, A	243

J

James, F. L	196
James, J. F	196
Jameson, R	215
Jamieson, T. F	215
Jeffreys, J. G	210, 215, 292
Johnson, H. A	196
Joly, N	243, 244
Jones, F. W. O. Rymer	292
Jones, T. Rupert	184, 186, 196, 201, 202, 208, 211, 215, 216, 217, 221, 222, 231, 243, 271, 292, 293
József-től, S	266
Judd, J. W	217
Julien, A. A	184

K

	PAGE.
Kanmacher, F.	217
Karrer, F.	196, 262, 267, 268, 292
Karsten, H.	268
Kaufmann, F. J.	268
Keeping, W.	217
Keferstein, C.	268
Kent, W. S.	217
Keyserling, C.	287
Kinahan, G. H.	217
King, W.	184, 185, 207, 217, 218
Kirkby, J. W.	214, 218, 222
Klein, J. T.	268
Knorr	283
Koch, A.	268
Koch, F.	268
Koch, F. K. L.	268
Kölliker, A.	268
Kübler, J.	268, 269

L

Lachmann	236, 269
Lafont, A.	269
Lamarck, J. B. de	243, 244
Lamplugh, G. W.	218
Lankester, E. R.	218, 269
Latham, A. G.	218
Lartet, L.	292, 293
Latreille, P. A.	244
Laube, G	185
Lea, I.	185, 196
Lebour, G. A.	218
Ledermüller, M. F.	269
Legg, M. S.	218
Leidy, J.	185, 196
Lesser, R.	266
Leymerie, A.	243, 244, 245
Linnaeus, C.	269
Linnaeus, C. A.	269
Linton, J.	218
Lister, M.	218
Liversidge, A.	218
Locard, A.	245
Logan, W. E.	185, 186
Lomnicki, M.	269
Lorić, (Dr.) J.	269
Lory, C	245
Lovisato, D.	245
Lyell, C.	196, 197, 245

M

	PAGE.
MacCoy, F.	218
Macdonald, J. D.	218, 293
Macgillvray, W.	219
Mackie, S. J.	219
Maffit	192
Maitland, R. T.	269
Mantell, G. A.	219, 293
Mantell, W.	293
Manton, W. G.	219
Marck, W. v. d.	292
Mareschkowsky, K. S	287
Manzoni, A.	242, 245
Marsson, T	269
Martens	269
Martin, K.	269, 293
Martini, F. H. W	269
Martonfi, L. A.	270
Martonfi, L	269
Massolongo	245
Maury, M. F.	197
Mayer, K	270
McAndrew, R.	219
M'Coy, T	219
Measures, J. W.	219
Meek, F. B.	197
Meneghini, G	248
Menke, C. F.	245
Merian, P.	270
Meyer, O	197
Michaud, A. L. P	247
Michelotti, G.	245
Miller, H. J.	270, 280
Milne-Edwards, A	245
Millet, F. W.	205
Mivart, (Sr.) G.	219
Möbius, K	186, 293
Moebius, K.	186, 270
Moggridge, M	242
Moll, J. P. C.	261
Möller, V. v.	270, 287
Montagu, G.	220
Montfort, D.	246
Moore, C.	219
Moore, J. C	202
Morris, J	219
Mortillet, G. de	246
Morton, S. G.	197
Moseley, H. N	219, 220

	PAGE.
Müller, J	270
Munier-Chalmas	220, 246
Murchison, R. I	245, 287
Murie, J	186
Murray, A	270
Murray, J	197, 220, 227

N

Needham, T. V	220
Neugeboren, J. L	270, 271
Nevill, T. H	220
Nicholson, H. A	186, 220
Nicolis, E	246
Niedzwiedzki, J	271
Nilsson, S	271
Norman, A. M	220
Northampton (Marquis of)	221
N. J	271
Nyst, H	271

O

Olszewski, (Dr.) St	271
Owen, D. D	198
Owen, S. R. I	221

P

Packard, (Jr.) A. S	198
Pareto, L	246
Parfitt, E	221
Parker, W. K	186, 196, 202, 208, 211, 216, 217, 221, 222, 243, 271, 293
Parkinson, J	222
Pasteur, J. D	276
Paul, K. M	256, 272
Peach, C. W	223
Pennant, T	223
Perry, G	223
Perry, J	223
Perry, J. B	186
Peters, K. F	272
Philippi, R. A	246, 272
Phillips, J	223
Pictet, F. J	247, 271
Planchus, J	247
Plancus, J	247
Potiez, V. L. V	247
Pourtales, L. F. de	198
Pratt, S. P	247
Prestwich, J	223
Price, F. G. H	223

	PAGE
Pritchard, A.	223
Pulteney, R.	223
Pusch, Geo. G.	272
Pusyrewski, (Prof.) P.	186

Q

| Quekett, J. | 219 |

R

Rackett, (Rev.) T.	219
Raulin, V.	247
Reade, J. B.	224, 247
Reade, T. M.	186
Reichert, C. B.	272
Reinsch, P. F.	272
Renevier, E.	243, 247
Reuss, A. E.	198, 224, 272, 273, 274, 275
Richeter, R.	275
Richthofen, Bron. V.	293
Richthofen, Fv. F.	293
Risso, J. A.	247
Robertson, D.	205, 213, 224, 226, 258
Roemer, F.	198, 293
Roemer, F. A	275, 276
Rogers, H. D	224
Roissy, F.	246
Rolle, F.	276
Rouault, A.	247
Rouillier.	287
Rousseau.	287
Rowney, T. H.	184, 185, 186, 218
Russegger, M.	293
Rütimeyer, L.	248, 276
Rutot, A.	276, 281
Ryder, J. A.	199

S

Sage, F. G.	248
Saint Fond, B. F.	276
Salter, J. W.	199, 225
Sandahl, O.	276
Sandberger, F.	248, 276
Sander Rang, A	248
Sandford, (Mr.)	186
Sars, G. O.	276
Sars, M.	276
Saussure, H. B. de.	248
Savi, P.	248
Schafhaeutl, C. E. v.	277

STATE GEOLOGIST. 307

	PAGE.
Schafhaeutl, K.	277
Schafhaeutl, K. E. v.	277
Schardt, A.	248
Schlicht, E. v.	277
Schlotheim, E. F. v.	277
Schlumberger, C.	199, 220, 225, 246, 248
Schlüter, C.	277
Schmarda	277
Schmelck, L.	277
Schmid, E. E.	277
Schneider, A.	248, 277
Schomburgh, R. H.	202
Schreiber, K. v.	277
Schroeter, J. S.	277
Schultze, M.	278
Schultze, M. S.	186, 225, 278
Schulze, F. E.	278
Schwager, C.	248, 278, 293, 294
Schweigger, A. F.	279
Scortegagna, F. O.	248, 249
Sequenza, G.	225, 249
Shacko, G.	276
Shone, W.	225
Shumard, B. F.	199
Siddall, J. D.	225
Siebold, C. T. E. v.	279
Silvestri, O.	249
Sinzo, J.	268
Sismonda, A.	249
Sismonda, E.	249
Six, Ach.	249
Smith, E. A.	199
Smith J. T.	225
Soldani, A.	250
Sollas, W. J.	225, 226
Sorby, H. C.	226
Sowerby, G. B.	226
Sowerby, J. de C.	294
Spencer, J. W.	199
Spengler, L.	279
Speyer, O.	279
Spratt, T.	287
Stache, G.	250, 279, 294
Steinmann, G.	279
Stewardson, G.	226
Stewart, S. A.	226
St. John, O. H.	199
Stöhr, E.	250
Stoliczka, F.	294

Strickland, H. E. 226
Studer, T. 250
Stur, D. 280
Stur D. v. 279
Suess, E. 250

T

Tallavignes. 250
Taranek, K. J. 280
Tate, R. 226
Tchihatcheff, P. de 250
Terrigi, G. 252
Terquem, O. 250, 251, 252, 268
Thomas, B. W. 196, 200
Thompson, W. 226, 227
Thomson, W. 187, 226
Thorpe, C. 227
Thurmann, J. 280
Tietze, (Dr.) E. V. 280
Tizard. 227
Toula, F. 280
Tournouer, (M) R. 252, 253
Tozzetti, G. T. 253
Turton, W 227
Tute, J. S. 227

U

Uhlig, V 280

V

Vander Broeck, E 253, 270, 271, 280, 281, 294
Van Broeck, E. 202
Vasseur, G. 253
Verbeek, R. D. M. 281, 294
Verneuil, E. de 199, 253, 287
Verneuil, E. P. de. 294
Verrill, A. E. 199
Vilanova, Y. P. J. 187
Villa, C. G. B. 253
Vincent, G. 281
Vine, G. R 227
Von Alberti, F. 253
Von Daday, E. 281
Von der Marck, (Dr.). 281, 292
Von Dunikowski, E 281, 282
Von Fritsch, K. 294
Von Hagenow, A. E. 282

	PAGE.
Von Hantken, M	253, 282
Von Keyserling, G. A.	287
Von Madarasz, S. E.	282
Von Mereschkowsky, C.	287
Von Robz, Z.	282
Von Schauroth, K. F.	282
Von Schlotheim, E. F.	282
Vorce, C. M.	199
Vosinaky	287

W

Wadsworth, M. E.	187
Walch, J. E. J	282, 283
Walford, E. A.	227
Walker, G.	227
Waller, E.	227
Wallich, G. C.	199, 227, 228, 229
Waters, A. W.	253
Weaver, T	229, 283
Webb, P.	290
Wetherell, N. T.	229
Whitaker, W.	229
White, C. A.	199
Whiteaves, J. F.	200
Whitfield, R. P.	200
Whitney, J. D.	187
Wichmann	293
Wilson, E.	229
Williamson, W. C.	229, 230
Winchell, N. H.	187
Winter, G.	283
Wolf, H. v.	283
Wood, J G.	230
Wood, J. E. T	294
Wood, W.	230
Woodward, A.	200
Worthen, A H.	197
Wright, E. P.	230
Wright, J.	205, 230
Wright, T. S.	230, 231
Wrisberg	283
Wyatt, J.	231

Y

Young, J	205, 231

Z

	PAGE.
Zborzewski, A ...	287
Zigno, A. de	253
Zittel, (Dr.)	283
Zittel, K. A	283
Zsigmondy, W	283
Zwingli, H	268, 269

www.ingramcontent.com/pod-product-compliance
Lightning Source LLC
Chambersburg PA
CBHW020053170426
43199CB00009B/274